Praise for *Sun in a Bottle*

"Substantive and lively . . . Seife writes with effortless clarity."
—*The New York Times Book Review*

"Award-winning science journalist Seife takes a long, hard look at nuclear fusion and the failure of one scheme after another to turn it into a sustainable energy source. . . . This approachable book should interest everyone concerned about finding alternative energy sources."
—*Publishers Weekly*

"With great explanatory skill, Seife explains how fusion works and why it is so hard to get power out of it. . . . Enjoyable . . . a must-read for anyone who wants to know the story behind ongoing multibillion-dollar attempts at bottling up the sun."
—*Science News*

"This book, like Mr. Seife's previous ones, is a model of what scientific history should be. It does not fudge or oversimplify facts. It does not talk down to its readers. What it does do is explain crucial scientific principles and complex debates over research in a scintillating, clear way that anyone can understand."
—*The Washington Times*

ABOUT THE AUTHOR

Charles Seife is the author of *Decoding the Universe*, *Alpha & Omega*, and *Zero*, which won the PEN/ Martha Albrand Award for first nonfiction book and was named a *New York Times* Notable Book. An associate professor of journalism at New York University, he has written for *Science* magazine, *New Scientist*, *Scientific American*, *The Sciences*, *Wired*, *The Economist*, and many other publications. He lives in New York City.

SUN

IN A BOTTLE

The STRANGE HISTORY of FUSION
and the SCIENCE of WISHFUL THINKING

Charles Seife

Penguin Books

PENGUIN BOOKS

Published by the Penguin Group

Penguin Group (USA) Inc., 375 Hudson Street, New York, New York 10014, U.S.A.
Penguin Group (Canada), 90 Eglinton Avenue East, Suite 700, Toronto,
Ontario, Canada M4P 2Y3 (a division of Pearson Penguin Canada Inc.)
Penguin Books Ltd, 80 Strand, London WC2R 0RL, England
Penguin Ireland, 25 St Stephen's Green, Dublin 2, Ireland (a division of Penguin Books Ltd)
Penguin Group (Australia), 250 Camberwell Road, Camberwell,
Victoria 3124, Australia (a division of Pearson Australia Group Pty Ltd)
Penguin Books India Pvt Ltd, 11 Community Centre, Panchsheel Park, New Delhi – 110 017, India
Penguin Group (NZ), 67 Apollo Drive, Rosedale, North Shore 0632,
New Zealand (a division of Pearson New Zealand Ltd)
Penguin Books (South Africa) (Pty) Ltd, 24 Sturdee Avenue,
Rosebank, Johannesburg 2196, South Africa

Penguin Books Ltd, Registered Offices:
80 Strand, London WC2R 0RL, England

First published in the United States of America by Viking Penguin,
a member of Penguin Group (USA) Inc. 2008
Published in Penguin Books 2009

1   3   5   7   9   10   8   6   4   2

Insert illustration credits
Page 1 (two images), 4 (top), 6 (top left and bottom): United States Department of Energy • 2 (left),
3, 6 (top right): Lawrence Livermore National Laboratory • 2 (right): DigitalGlobe • 4 (bottom):
Federation of American Scientists • 5 (top and bottom left): AP/Wide World Photos • 5 (bottom
right): Lynn Freeny/United States Department of Energy • 7: ITER • 8: Randy Montoya

THE LIBRARY OF CONGRESS HAS CATALOGED THE HARDCOVER EDITION AS FOLLOWS:
Seife, Charles.
Sun in a bottle : the strange history of fusion and the science of wishful thinking / Charles Seife.
p.   cm.
Includes bibliographical references and index.
ISBN 978-0-670-02033-1 (hc.)
ISBN 978-0-14-311634-9 (pbk.)
1. Nuclear fusion—History.   I. Title.
QC791.S45 2008
539.7'64—dc22   2008013135

Printed in the United States of America
Set in Legacy Serif ITC with ChollaSlab
Designed by Daniel Lagin
Illustrations by Thom Graves

CONTENTS

INTRODUCTION

Circe warned me to shun the island of the blessed sun-god,
for it was here, she said, that our worst danger would lie.

—*THE ODYSSEY*, TRANSLATED BY SAMUEL BUTLER

The dream is as ancient as humanity: unlimited power. It has driven generation after generation of scientists to the brink of insanity.

In 1905, after centuries of attempts to build perpetual motion machines, scientists discovered an essentially limitless source of energy. With his famous equation, $E = mc^2$, Albert Einstein discovered that a minuscule chunk of mass could, theoretically, be converted into an enormous amount of energy. Indeed, $E = mc^2$ is the equation that describes why the sun shines; at its core, the sun is constantly converting matter to energy in a reaction known as fusion. If scientists could do the same thing on Earth—if they could convert matter into energy with a controlled fusion reaction—scientists could satisfy humanity's energy needs until the end of time.

For the past half century, legions of physicists have been trying desperately to create a tiny sun in a bottle, trying to bring the stellar power of fusion to Earth. The quest for fusion is the story of scientists weaving an increasingly tangled web of secret, crazy, and brilliant schemes to harness the power of the sun. They are caught up in a complex tale that includes classified government experiments, billion-dollar scientific projects, and byzantine conspiracy theories. The quest for fusion is a tale of genius physicists who have changed the world forever—for better and for worse—and of secret-spilling whistleblowers, jealous researchers, brilliant tinkerers, and backstabbing politicians.

The stakes are enormous—and they are getting higher by the day. The world's supply of oil is no longer assured to meet humanity's energy needs; worse yet, the threat of global warming is forcing governments to find sources of power other than fossil fuels. In the long term, fusion is the only option. Humanity will suffer if researchers don't solve its problems.

Scientists have broken under the pressure. Others have been forced to make a heartwrenching decision to give up their dreams and disavow their work or to be driven from the fold of mainstream science. Over and over again, the dream of fusion energy has driven scientists to lie, to break their promises, and to deceive their peers. Fusion can bring even the best physicists to the brink of the abyss. Not all of them return.

CHAPTER 1

THE SWORD OF MICHAEL

He took not away the pillar of the cloud by day, nor the pillar of fire by night, from before the people.

—EXODUS 13:22

The fires were still burning over Hiroshima, the charred and faceless victims were still slouching toward Asano Park, when President Harry S. Truman told the world about a new weapon. "The force from which the sun draws its power has been loosed against those who brought war to the Far East," the announcement read. Mankind had unleashed unheard-of energy from deep within the atom and used it to destroy a city.

From the very beginning of the atomic age, Americans were enthralled and frightened by the prospect of this inconceivable power. By splitting uranium and plutonium atoms, scientists had made a weapon by using the very same principle that made the sun shine: $E = mc^2$.

The scientists who worked on the Manhattan Project, the super-secret program to build the first atom bomb, looked back on their achievement with a mix of awe and horror. To J. Robert Oppenheimer, the head of the Manhattan Project, the atom bomb represented a loss of innocence, a fall from grace that could mark the end of civilization. Others, however, such as the Manhattan Project physicist Edward

Teller, saw that the atom bomb was just the beginning of a nuclear arms race. And just over the horizon, Teller realized, was a much greater weapon than even the atom bomb, one thousands of times more powerful.

This new weapon, the "Super," would unleash a power not yet seen on Earth: fusion. Instead of breaking atoms apart to release energy (*fission*), the superbomb would stick them together (*fusion*) and release even more. While this might seem to be a subtle difference, fusion, unlike fission, had the potential to produce weapons of truly unlimited power. A single Super would be able to wipe out even the largest city—a task far beyond even the bombs of Hiroshima and Nagasaki. A fusion bomb would be the ultimate weapon.

It would also split the scientific community in two and would drive humanity to the brink of ruin. The quest to unleash the power of the sun upon the Earth had an inauspicious start, to say the least.

━━━━━━

The atom bombs that destroyed Hiroshima and Nagasaki were fission, not fusion, weapons. Fission and fusion are siblings. Both get their power from converting the mass at the heart of the atom into energy.

Scientists got their first taste of that power in 1898, when the husband-and-wife team of Pierre and Marie Curie discovered a substance with a curious property. Radium, as they called it, seemed to produce energy from nothing. This was, of course, impossible. The most rigid laws of physics, the laws of thermodynamics, seemed to forbid the spontaneous creation of energy. But the Curies were quite certain of what they were observing. A hunk of radium constantly produced heat like a little oven; every hour, a chunk of radium generated enough heat to melt its own weight in ice. It would do this, hour after hour, day after day, and year after year. No chemical reaction could possibly sustain itself for so long and generate so much energy. Whenever the Curies cooled a piece of radium, it would heat itself back up. Indeed, the radium would always be hotter than its surroundings, even though there were no external sources of heat. Marie Curie herself was baffled.

She suspected that some sort of change was happening at the center of the radium atom, but she didn't know what it could be—or how such a tiny chunk of matter could produce so much energy.

The answer would come a few years later when the young Albert Einstein formulated his theory of relativity. The theory revolutionized the way scientists perceive space, time, and motion. One of the equations that came out of the theory was $E = mc^2$, the most famous scientific equation of all time. $E = mc^2$ showed that matter, m, could be converted into energy, E. This was the secret to the seemingly endless fountain of energy coming from radium.

If you put a gram of radium in a sealed ampule, over many, many years the radium (a whitish metal) will gradually disappear. In fact, the atoms of radium spontaneously split apart and vanish from view. But they don't disappear entirely. When an atom of radium breaks apart, it tends to split into two smaller pieces. The heavier of the two is a gas known as radon; the lighter is helium, and the Curies detected both helium and radon emanating from their radium sample.

Radium—a big heavy atom—breaks up into helium and radon, and when scientists looked carefully at the weights of those atoms, they realized the source of the heat. Some of the mass of the radium was missing. If you add up the mass of one atom of radon and one atom of helium, they make up 99.997 percent of the mass of the radium atom from which both sprang. The other 0.003 percent simply vanishes. When radium breaks apart, the parts are lighter than the original atom.

Here was the answer to the puzzle of excess energy. The whole atom weighed more than the sum of the parts. When the radium atom spontaneously broke apart, some of its mass changed into energy, just as Einstein's equation allows. The m had become E. The missing mass was only a tiny fraction of what made up the atom, but even tiny chunks of mass are converted into enormous amounts of energy. It was energy on a scale much, much greater than humans had ever accessed before.

As World War II loomed, scientists began to realize that this energy could become a potent weapon. Less than a month before Germany invaded Poland in 1939, Einstein warned President Franklin Delano

Roosevelt of the possibility of a bomb made from uranium, a metal that, like radium, releases energy when it breaks into pieces. Such a bomb would be extremely powerful—and there were ominous signs that the Nazis were already on their way to building one. For example, Germany had halted the uranium trade in occupied Czechoslovakia.

Uranium—in particular, one variety known as uranium-235—is an ideal material for a weapon. Its atoms are very sensitive; hit one with a subatomic particle and it fissions into fragments. Unlike decaying radium, which tends to cleave cleanly into two parts, a fissioning uranium atom shivers into a number of smaller chunks, including a handful of neutral particles known as neutrons. These neutrons then fly away from the shattered atom.

In a vacuum, the neutrons continue merrily on their way without bumping into anything else. However, a chunk of uranium is not a vacuum; it is a space crowded full of billions and billions of other uranium atoms. Once a single atom splits apart, within a tiny fraction of a second the resulting neutrons might slam into two or three other uranium atoms. These collisions cause those atoms to split, and in the process, each releases two or three more neutrons. All these neutrons slam into other atoms, splitting them, releasing even more neutrons. If the conditions are right—if enough uranium atoms are in a small enough space—then the process snowballs out of control in less than a blink of an eye. One atom fissions, and its neutrons cause two more to split. These cause four more to fission, causing eight to break apart, then sixteen, thirty-two, sixty-four, and so forth. After ten rounds, over two thousand atoms have split, releasing neutrons and energy. After twenty rounds, it's more than two million atoms; after thirty rounds, two billion; after forty, more than a trillion. This is a chain reaction.

A chain reaction, if it gets big enough, can level a city. Every time a uranium nucleus splits, it releases energy. Like radium, a uranium atom loses mass when it splits. In a tiny instant, the mass is converted into energy, just as $E = mc^2$ predicts. The more atoms that split in the chain reaction, the more energy is released. After forty rounds of splitting

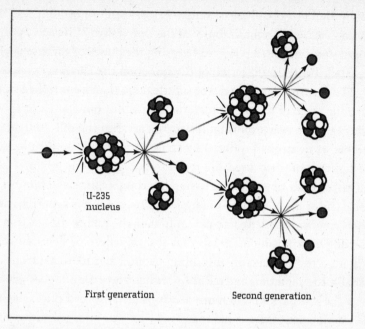

U-235
nucleus

First generation   Second generation

FISSION CHAIN REACTION: When a neutron strikes a U-235 nucleus, the nucleus splits, releasing more neutrons, which strike more nuclei, and so on.

uranium atoms, the energy is roughly enough to light an incandescent lightbulb for about a second. After eighty rounds, a mere fraction of a second after the chain reaction begins, the result is more energetic than the explosion of ten thousand tons of TNT, roughly the size of the blast that eventually destroyed Hiroshima.

In 1939, though, the idea of fission—and a chain reaction that would release a tremendous amount of energy—was just a theory. Before World War II began, scientists were uncertain whether the theory was right—and if so, how to turn that theory into the hard reality of a useful weapon. It took two years of cogitation and experimentation for the consensus to build: it was possible to build a powerful bomb out of uranium-235 or plutonium-239 (an atom created in the lab by bombarding uranium with neutrons). Nuclear theory progressed quite rapidly; by 1942, the physicist Enrico Fermi was busy building the first nuclear reactor in a squash court* at the University of Chicago. Fermi's project was a major step toward releasing the power of the atom—and eventually bringing the wrath of the sun upon the Earth.

The core of a nuclear reactor is little more than a controlled chain reaction: a pile of fissioning material that is not quite at the stage of entering a runaway explosion. Scientists arrange the pile so that the number of neutrons produced by splitting atoms is almost precisely the right amount to keep the reaction going without getting faster and faster; each generation of fission has roughly the same number of atoms fissioning as the last. In physics terms, the pile is kept right near *critical* condition. Scientists can manipulate the rate of the reaction by inserting or removing materials that absorb, reflect, or slow neutrons. Pull out a rod of neutron-absorbing material and more neutrons are available to split atoms and release more neutrons: the pile goes critical. Drop the rod back in and more neutrons are absorbed than released: the reaction sputters to a halt.

At 3:36 PM on December 2, 1942, Fermi and his colleagues pulled a neutron-absorbing rod out of a pile of graphite and uranium oxide. The radiation counters chattered. Fermi had created the first self-sustained nuclear reaction. The pile had gone beyond critical; more neutrons were being produced by each generation of fission than the last. The reactor was producing more and more and more energy. About a half hour later, Fermi ordered the control rods back into the pile, and the reaction stopped. At its peak the reactor was producing about half a watt of power, almost enough to light a dim Christmas-tree lightbulb. Nevertheless, the possibilities were enormous: Fermi's reactor showed that nuclear power could, in theory, light up a city. Or destroy it.

It was for the latter purpose that the Manhattan Project was born. At its head was a quirky and difficult scientist, J. Robert Oppenheimer, a man who would achieve fame through fission and be destroyed by fusion.

---

* Apparently, Soviet translators erred when they received word of Fermi's experiment. Soviet scientists were led to believe that the nuclear pile was in a "pumpkin field" instead of a "squash court."

Oppenheimer was not an obvious choice to lead America's race to build an atom bomb. He was a good physicist, but he was a theorist—and the Manhattan Project was, fundamentally, an engineering project. Oppenheimer was about as far from the stereotypical get-your-hands-dirty engineer as possible.

The aristocratic Oppenheimer grew up in a wealthy family, but what was particularly striking about him was his quick mind. He mastered more than half a dozen languages, including Sanskrit. He was an adept theoretician but struggled with the more practical side of science; he had difficulty even with basic tasks such as soldering copper wires. After graduating from Harvard, he went to Cambridge in England to work in the lab of the famous experimentalist J. J. Thomson. There, the already high-strung Oppenheimer became unglued.

Oppenheimer had a difficult time at Cambridge; in his mind, his experiments were failures, and he contemplated suicide. He also contemplated murder. In 1925, he suddenly tried to strangle a childhood friend, and his behavior got even more bizarre from there. On a vacation in Corsica with two friends, he abruptly announced, "I've done a terrible thing." He said that he had poisoned an apple and put it on the desk of another brilliant physicist at Cambridge, Patrick Blackett. When everyone got back to the university, they found out that Blackett was unharmed, and Oppenheimer's friends were left wondering whether the apple was real or just a figment of Oppenheimer's feverish imagination.

The bizarre behavior became less acute once Oppenheimer relocated to the University of Göttingen in Germany. In the 1920s, Germany was the world leader in theoretical physics—home to Einstein, Max Planck, Werner Heisenberg, Max Born, and many of the other leading lights of the day—and Oppenheimer established himself as a brilliant young physicist. However, he was still depressive. He was also vain, arrogant, and occasionally nasty. He had a habit of making people feel small and insignificant; he detested his "beastliness" but was unable to control it. Nevertheless, soon after moving back to the United States to become a professor at the University of California at Berkeley, he acquired a circle of devotees thanks to his brilliance and wit.

Despite Oppenheimer's prickliness, everyone—even the occasional general—was impressed with the young professor. "He's a genius," wrote General Leslie Groves, the military head of the Manhattan Project and the man who chose Oppenheimer to lead the scientific effort. "Why, Oppenheimer knows about everything. He can talk to you about anything you bring up. Well, not exactly. I guess there are a few things he doesn't know about. He doesn't know anything about sports." This was by no means the most serious of his flaws, as far as the military was concerned.

Oppenheimer was a security risk—he was absolutely surrounded by Communists. His brother and sister-in-law were members of the Communist Party. His first fiancée, Jean Tatlock, had been a member, too. His wife Kitty's first husband had been an official in the party and had been killed fighting on the leftist side during the Spanish Civil War. The army knew about all these connections, yet Groves insisted that Oppenheimer lead the most sensitive military project of World War II. In October 1942, Oppenheimer accepted his new post and began assembling the biggest scientific project in the history of mankind.

Laboratories devoted to the atom bomb effort sprang up around the country. Los Alamos, perched on a mesa in the New Mexico desert, was the intellectual heart of the Manhattan Project. Other facilities, such as one at Oak Ridge in Tennessee and another at Hanford in Washington, were crucial to figuring out the best way to separate bombworthy uranium-235 from the much more common uranium-238 and how to manufacture plutonium-239.* However, the big minds roamed at Los Alamos: Oppenheimer, Hans Bethe, Richard Feynman, Stanislaw Ulam, John von Neumann, Enrico Fermi, and Edward Teller.

Teller, a Hungarian émigré and, arguably, a better theoretician than

---

* In the interest of secrecy, Manhattan Project scientists seldom referred to these compounds by their real name. Uranium-235 was referred to by the code names "magnesium" or "25"; plutonium-239 was "copper" or "49."

Oppenheimer, was brought to the University of Chicago in mid-1942 by the Manhattan Project just as it was getting under way. When Teller arrived, nobody assigned him a task, so he set to work trying to design the ultimate weapon, more powerful even than the one the project's scientists were trying to build. He envisioned a superbomb that used fusion instead of fission. If it worked, it would dwarf an atom bomb just as surely as an atom bomb would dwarf conventional explosives. Teller became obsessed with wielding the power of the sun. It was an obsession that molded him into one of the darkest and most twisted figures of American science. "He's a danger to all that's important," said his fellow physicist Hans Bethe. "I really do feel it would have been a better world without Teller."

───

Teller was born in Budapest, the child of a successful lawyer. In 1919, when he was eleven years old, the Communist Béla Kun swept to power and declared Hungary a Soviet state. "The communists overturned every aspect of society and the economy," Teller later wrote. "My father could no longer practice law." Two soldiers moved into the Tellers' home, and young Edward came to know hunger. "There was no food (or any other kind of goods) for sale in the stores now owned by the communists....As I recall, cabbage was often all we could find. I still dislike cabbage."

After rampant inflation, a coup attempt, a purge, and a military defeat, Kun's regime ended before the year was out. But the whole experience left Teller with an almost monomaniacal hatred of Communism. In large part, his actions over the next few decades—his attempt to build an arsenal of unlimited power—would be driven by that hatred.*

Thus Teller's vision of a superweapon was possible because there is more than one way to extract energy from the atom. Fission is the easy

---

* Teller's obsessive hatred of Communists and single-minded desire to build fusion weapons reportedly led Enrico Fermi to tell him, "In my acquaintance, you are the only monomaniac with several manias."

way. Just get enough fissile material (such as uranium-235 or plutonium-239) in a small enough space and a chain reaction will start on its own. Heavy atoms will split into fragments, converting mass into energy and creating an enormous explosion. The main problem is getting that fissile material. Neither uranium-235 nor plutonium-239 was easy to obtain, especially with the state of knowledge in 1942 and 1943.

Fusion is another way to convert mass into energy; it's the opposite of fission. In fission, heavy atoms split and the sum of the parts is lighter than the original atoms. In fusion, light atoms stick together, and the whole resulting atom is lighter than the sum of the parts that made it. The missing matter—the stuff that disappears when the light atoms combine—becomes energy.

Fusion is several times more powerful than fission; more of the mass of each reacting atom is converted into energy. Better yet, it is much easier to find the fuel for fusion—light atoms like hydrogen—than it is to find the uranium or plutonium fuel for fission. The oceans are filled with hydrogen's heavier sibling, deuterium, a great fuel for fusion reactions. It's not terribly difficult to extract a practically unlimited amount of the stuff.

Of course, there is a downside. The fusion reaction is extremely difficult to start, and even harder to keep going long enough to produce large quantities of energy. Atoms tend to repel each other, so it is very hard to get them close enough so that they stick together. You need an enormous amount of energy to slam two atoms together forcefully enough to overcome that repulsion and get them to fuse.

For a fission reaction, you just need to get a lump of uranium big enough. For fusion, you need to manipulate your fuel in some tricky ways. First, you've got to compress the fuel into a tiny parcel. This keeps the atoms in close proximity to one another (so they have a chance of colliding). That, in itself, is not so hard; the trick is to keep the atoms very hot as well. Only at tens or hundreds of millions of degrees are the atoms moving fast enough to have a chance of fusing when they do collide. When you heat something, it expands—the atoms try to escape in all directions. Thus, it is very hard to keep a very hot thing compressed very tightly. So, the basic problem in fusion is that it is very difficult to heat something to the right temperature and, simultaneously, keep the atoms close enough together. Without both things working concurrently, a fusion reaction won't get going.

Making matters worse, if you are lucky enough to start a fusion reaction, your own success works against you. When the fusing atoms release energy, they pour heat into their surroundings. This makes the neighboring atoms hotter. The hotter the atoms get, the more the fuel expands and the harder the atoms try to escape. The packet of fuel attempts to blow itself apart. Unless the conditions are just right, a fusion reaction will snuff itself out before it produces any appreciable energy.

Nevertheless, if scientists could get a fusion reaction going even for a few fractions of a second, its power would be virtually limitless. It could be much, much more deadly than a mere fission bomb.

This is the idea that obsessed Teller soon after he arrived in Chicago. Unlike most of his colleagues, he was not terribly interested in working on the fission bomb. In his mind, the theoretical problems had already been solved, so he spent his energy trying to come up with even better weapons: fusion bombs. Within a month of his arrival, Teller had not only concluded that it was possible to create a fusion bomb that would dwarf anything the Manhattan Project would be able to offer, but had also convinced himself that he knew precisely how to build one. It would be years before he figured out how wrong he was.

In 1942, though, Teller, full of enthusiasm, brought the idea to the attention of his colleagues. They quickly dubbed the new weapon the Super. By August, he and his fellow physicists were giving astounding estimates of the destructive power of a Super-like fusion weapon. A report at the time estimated that one would blow up with the energy of one hundred megatons of TNT, about seven thousand times bigger than the eventual size of the Hiroshima bomb. Teller, a tremendous optimist,* was convinced that fusion was easy.

---

* Teller was so ridiculously optimistic that fellow physicists measured enthusiasm in "Tellers" just as they would measure mass in kilograms or time in seconds.

Once you have an atom bomb, he argued, you can dump the enormous power of an exploding atomic weapon into a tank of deuterium—heavy hydrogen. The hydrogen, heated to millions of degrees, would begin to fuse and generate energy in a *thermonuclear* reaction. This was essentially the idea behind Teller's Super: it was, more or less, an atom bomb at one end of a vessel full of heavy hydrogen. The exploding bomb would trigger a wave of fusion in the vessel. If it worked, Teller argued, this Super had unlimited capacity for destruction.*

To Teller, the easy part was building a weapon of tremendous power. The hard part was building a weapon that would *not* be so destructive that it would kill everybody on Earth. In Teller's fertile imagination, an atom bomb that ignited a tank of hydrogen might ignite the air itself. (The nitrogen that makes up 80 percent of the atmosphere is a light atom, and just like hydrogen it will fuse if the conditions are right.) Teller's initial calculations showed that an atomic explosion might induce nitrogen atoms in the air to fuse with each other. The runaway explosion would quickly destroy the world in a gigantic nuclear furnace—even the weak Manhattan Project bomb might mean the end of life on Earth. When Hans Bethe double-checked Teller's assumptions, though, he found reason to relax. "I very soon found some unjustified assumptions in Teller's calculation that made such a result extremely unlikely, to say the least." If a fusion reaction got going, there was too much energy lost through radiation to get the atmosphere hot enough to cause a chain reaction of fusing nitrogen.† The

---

* The Los Alamos physicist Robert Serber later wrote, "On Edward Teller's blackboard at Los Alamos I once saw a list of weapons—ideas for weapons—with their abilities and properties displayed. For the last one on the list, the largest, the method of delivery was listed as 'Backyard.' Since that particular design would probably kill everyone on Earth, there was no use carting it elsewhere."

† That didn't end the speculation. As General Groves later recounted, "I had become a bit annoyed with Fermi the evening before [the first atomic bomb test], when he suddenly offered to take wagers from his fellow scientists on whether or not the bomb would ignite the atmosphere, and if so, whether it would merely destroy New Mexico or destroy the world."

world was safe. Fusion was much more difficult than Teller initially imagined.

Fusion was so hard, in fact, that the Super, at least as originally designed by Teller, wouldn't work at all. According to the physicist Robert Serber, "Edward first thought it was a cinch. Bethe, playing his usual role, knocked it to pieces." Hans Bethe showed that the fireball in Teller's Super device would cool very rapidly. Here, too, the energy of a budding fusion reaction would quickly drain away through radiation; the fusion would snuff itself out before it really got going. It wasn't an insurmountable obstacle, but it was enough of a problem for the Manhattan Project physicists to put Teller's idea on the back burner. In 1943, a review committee decided that all the lines of research for the project—and for its theoretical physics division, which had relocated to Los Alamos—were worthwhile except for one: fusion. Instead of trying to build superweapons, the committee argued, the lab must concentrate its efforts on building atomic weapons to end the war.

Teller was disappointed that his pet project was stalled. Bruising his ego further, Oppenheimer appointed Bethe to be the head of the theoretical physics division. Teller thought the appointment would be his—and he apparently took both slights personally.

This was the turning point in Teller's career. It was at this moment that Teller, the brilliant physicist, started becoming defined by his character flaws: his egocentrism, his nearly manic optimism, and his paranoia. All these traits would play a role in the coming tragedy, but it was the paranoia that led Teller to blame a single individual for all the insults he received at the hands of the Manhattan Project. He was refused his rightful position as head of theory at Los Alamos, and the Super was mothballed all because of one man: J. Robert Oppenheimer.

Oppenheimer and Teller would soon become bitter enemies. The two were very different. Oppenheimer, gaunt and aristocratic, was quite unlike the limping, bushy-browed Teller.* The most striking

---

* Teller limped because of an accident in his youth. At the age of twenty, he jumped off a tram and nearly lost his right foot.

difference was their politics. Oppenheimer, a leftist who flirted with Communism, was bound to clash eventually with Teller, the rabid anti-Communist.

However, in July 1945 the Teller-Oppenheimer feud was yet to ignite. It was a triumphant time for both physicists. The Los Alamos scientists had nearly overcome all the technical problems that faced them; they had manufactured and machined enough plutonium to build a "gadget" named Jumbo and had built an intricate cage of explosives that would force all the metal to assemble into a critical mass and explode. The scientists began to wager about how big the first atomic explosion—Trinity—would be. Oppenheimer bet that it would be the equivalent of a mere three hundred tons of TNT. Teller, ever the optimist, guessed that it would be forty thousand tons. It was raining in the predawn hours the day of the test, yet Teller was sharing his bottle of sunscreen with his colleagues.

When the New Mexico desert suddenly erupted with a light brighter than the noonday sun, the Manhattan Project scientists were relieved and jubilant. When a similar flash erupted over Hiroshima, the feelings were much more somber. When the war ended with Japan's unconditional surrender, Oppenheimer, like many of his scientific colleagues, lost his taste for weapons work.

By mid-September, half the staff at Los Alamos was already gone. Oppenheimer stepped down a month later—and he was warning about the dangers of adding atomic weapons to the world's arsenal. "The time will come when mankind will curse the names of Los Alamos and Hiroshima," he prophesied while accepting a military award in November. Bethe's departure left Los Alamos without a head of theory, the very post that Teller coveted, and Teller was offered the job. But Teller would only accept if the lab would devote its resources to developing better bombs—most likely a fusion weapon. Alas, the lab was to turn its attention to production rather than to designing fusion weapons. "There was no backing for the thermonuclear work. No one was interested in developing a thermonuclear bomb," huffed Teller. "No one cared."

Los Alamos was dissolving around him, and few Manhattan Project scientists seemed interested in developing the fusion bomb. Teller decided to pack his bags and move back to the University of Chicago. His relationship with Los Alamos wasn't over, however. He would consult for the laboratory during the postwar years, and he would soon return to the New Mexico complex.

Teller's dream of unlimited power was just a little premature. In just a few years, the United States would embark on a crash effort to develop fusion weapons.

———

The decision to build fusion weapons came from paranoia and fear. Even though the Americans had a monopoly on nuclear bombs, there was the nagging worry that the Soviets would soon build their own atomic weapons. Once that happened, Teller reasoned, they would certainly invade—unless America had an even bigger weapon in its arsenal: the Super. "Edward offered to bet me that unless we went ahead with his Super," wrote a colleague, "he, Teller, would be a Russian prisoner of war in the United States within five years!"

Just after the war ended, Teller tried to get the Super program started again. At a conference in April 1946, Teller and two dozen key scientists met to discuss whether a superbomb was feasible, and if so, what its future should be. There is some debate as to what the conference participants actually concluded, but the report was sanguine: "It is likely that a super-bomb can be constructed and will work," it said, adding that if doubts about the design proved to be true, "simple modifications of the design will render the model feasible." The report reflected Teller's unflagging optimism. (After all, he wrote the thing.) He was promising that fusion was within reach.

In truth, though, the road to the superbomb would be harder than Teller imagined. Not only was his design flawed, but he also had to overcome political opposition. Oppenheimer and his cronies were trying to get the United States to give up its monopoly on atom bombs—by giving nuclear secrets to the Communists. To Teller, it was madness; it was almost treasonous.

In March 1946, the month before Teller's Super conference, Oppenheimer and a government committee made the radical suggestion that "inherently dangerous" activities such as mining uranium should be put under international control and that all nations, including the Soviet Union, should have access to nuclear knowledge. As idealistic as this scheme might seem, at least in retrospect, it became official U.S. policy within a few months. The United States' representative to the UN Atomic Energy Commission, Bernard Baruch, presented such a plan to the United Nations. It was "a choice between the quick and the dead," he told the world. "We must elect world peace or world destruction." Not all nations agreed with that simplistic dichotomy. The Soviet Union opposed the proposal, and by the end of 1946 the plan was dead. It soon became clear why.

On September 3, 1949, a modified B-29 bomber flying off the coast of the Kamchatka Peninsula picked up alarming traces of radiation. It was the first sign of a radioactive cloud that soon drifted across the Pacific, the United States, and Canada before crossing the Atlantic and circling the world. Physicists around the United States scrambled to figure out the source of the radiation. It did not take long. The radioactive cloud had elements that showed that it was the result of nuclear fissions. It was fairly clear: the Russians had their own atom bomb. America's nuclear monopoly had ended much more quickly than anyone expected. On August 29, 1949, in the middle of the Kazakh steppe, a nuclear cloud had mushroomed to life. The Soviets called it "First Lightning"; the stunned Americans nicknamed the first Russian atom bomb test "Joe-1."

The timing could hardly have been worse. On September 21, Mao Tse-tung announced the formation of the People's Republic of China. A quarter of the world's population suddenly had a red flag flying above their heads. Two days later, President Truman had to announce the news of Joe-1. "We have evidence that within recent weeks an atomic explosion occurred in the U.S.S.R.," he said, and attempted to reassure a frightened nation. "Ever since atomic energy was first released by man, the eventual development of this new force by other nations was to be expected." Even so, the Russians had caught up with the Americans much more quickly than anticipated.* The hope of unilateral disarmament was gone forever. When Teller heard the news, he called Oppenheimer on the telephone, perhaps hoping to spur him to pursue fusion weapons. "Keep your shirt on!" was Oppenheimer's curt rejoinder.

It wasn't the response that Teller hoped for. He thought his Super provided an easy answer to the crisis. By developing a superbomb more powerful than even atomic weapons, the United States could keep its lead over the Soviet Union. But some scientists viewed pursuing fusion weapons as an inherently immoral act, one that might lead to the destruction of humanity. Physicists began to divide into two groups: pro- and anti-fusion.

Oppenheimer was in the latter camp. A month after Truman's announcement, Oppenheimer convened a meeting of a handful of scientists, politicians, and engineers—the General Advisory Committee (GAC)—who advised the new Atomic Energy Commission (AEC). The AEC had been formed in 1945 to oversee nuclear research, so the GAC, chaired by Oppenheimer, was extremely influential in setting the direction of America's nuclear weapons program. Oppenheimer's GAC was firmly anti-fusion. While the committee stated that it would be wise to increase the United States' capacity to build and research nuclear weapons, the GAC raised technical questions about the feasibility of the Super—and also made a firm statement about the morality of pursuing Teller's dream of weapons of unlimited power. "We are all agreed that it would be wrong at the present moment to commit ourselves to an all-out effort toward its development." Some scientists on the committee went further. Oppenheimer and five others railed against the morality of a fusion weapons program. "A super bomb might become a weapon of genocide," they wrote. "We believe a super

---

* This was, in part, because Russian military intelligence had penetrated the Manhattan Project. Klaus Fuchs, a physicist who was involved at the highest level of theoretical work on the atomic and hydrogen bombs, was a spy.

bomb should never be produced." To Enrico Fermi and fellow physicist Isidor Rabi, also on the committee, Oppenheimer's statement didn't go far enough. "The fact that no limits exist to the destructiveness of this weapon makes its very existence and the knowledge of its construction a danger to humanity as a whole," they argued. "It is necessarily an evil thing considered in any light." The GAC recommended shelving the Super project.

To Teller, who had returned to Los Alamos full-time, Joe-1 was the realization of his worst fears. A militant Communist state had just destroyed his home country; in August, Hungary officially declared itself Communist and became a satellite of the Soviet Union. And now the belligerent USSR was menacing his adopted home, the United States. Oppenheimer, with his Communist sympathies, was setting the United States on a course of self-destruction. Teller was certain that forsaking the Super, as Oppenheimer wanted, would ensure Soviet world domination within a few years. He had to thwart Oppenheimer's destructive influence. Teller started to make his case directly to Congress, and the Oppenheimer-Teller war began in earnest.

Teller had many allies also lobbying for the Super. A number of scientists and politicians agreed that an arms race with the Soviet Union was inevitable and thought that the Super was crucial to keeping the Soviets at bay. Lewis Strauss, a commissioner of the AEC, urged President Truman to launch a crash project to build a fusion weapon—even raising the specter that the Russians had taken the lead. The influential Berkeley physicists Luis Alvarez and Ernest Lawrence, too, stumped for a fusion bomb program. Congress was receptive to the fusion hawks' arguments. When Teller traveled to Washington, he quickly found an ally in Brien McMahon, the chairman of the Senate's Special Committee on Atomic Energy.

Even before the nuclear ash from the Joe-1 test had dispersed, the battles were under way. Hans Bethe, after being approached by Teller, initially agreed to work on the fusion bomb. Shortly after talking to Oppenheimer, however, Bethe backed out. Teller blamed Oppenheimer for the reversal. And those on either side of the divide—the pro-fusion and anti-fusion camps—began to distrust and dislike each other.

Teller and his allies looked upon Oppenheimer as an obstructionist and began to conclude that his actions damaged the military capability of the United States. Teller would later testify that Oppenheimer and his allies set back the fusion bomb effort by five years. The anti-fusion weapons side looked on with disgust as the hawks lobbied for the superweapon. For example, David Lilienthal, the first chairman of the AEC, was shaken by the "bloodthirsty" push for a fusion bomb. "The day has been filled, too, with talk about supers, single weapons capable of desolating a vast area," Lilienthal wrote in his journal in October 1949. "Ernest Lawrence and Luis Alvarez in here drooling over the same. Is this all we have to offer?" Teller's image, too, suffered, as he pressed harder and harder for the Super. "Now I began to see a distorted human being, petty, perhaps nearly paranoid in his hatred of the Russians, and jealous in personal relationships," wrote the Los Alamos physicist John Manley.

The scientists battled about whether or not to pursue fusion weapons, and the fight worked its way up to the president. Truman deliberated. Would he back the Super project or not? The pressures were building. Anti-Communist hysteria was sweeping the country, and the populace would clamor for a fusion bomb if they knew it existed.

They soon knew. On November 18, 1949, the *Washington Post* carried an alarming story on page 1. "[Scientists] are working and 'have made considerable progress' on 'what is known as a super-bomb' with '1000 times' the effect of the Nagasaki weapon," the article read. Soon, Truman was fielding questions at press conferences about the hydrogen bomb. The public clearly wanted a superweapon to counter the Soviet threat. The easy solution to the Russian problem—the Super—was becoming hard to resist. And then came the final blow.

On January 27, 1950, British police arrested Los Alamos physicist Klaus Fuchs, who confessed to being a spy. All of a sudden it became clear why the Russians were able to build an atom bomb so quickly. Worse yet, Fuchs had been involved in discussions about the fusion bomb; in fact, he was coholder of a key secret patent having to do with the method used to ignite the first working hydrogen bombs. The Russians knew all about the fusion bomb—and they had likely already begun research. Truman felt he had little choice.

Four days later, the president of the United States issued a public statement to his citizens. "It is part of my responsibility as Commander in Chief of the Armed Forces to see to it that our country is able to defend itself against any possible aggressor," it read. "Accordingly, I have directed the Atomic Energy Commission to continue its work on all forms of atomic weapons, including the so-called hydrogen or superbomb."

Truman's hand had been forced, but he had just made a dangerous decision. He had committed the United States to an arms race with the Soviet Union that would make both countries insecure and lead the world to the brink of destruction, all for the sake of a fusion weapon that, at the time, was merely a figment of Teller's fertile imagination.

Truman almost certainly didn't know this, but when he made his announcement, the fusion project at Los Alamos was entering its darkest time. Just weeks before, calculations from Los Alamos were starting to prove that Teller's fusion bomb was a flop.

It was hard—damnably hard—to get enough light atoms hot enough and dense enough to create a hydrogen bomb. Teller, as a theorist, made his best guesses as to how to engineer such a device and estimate what was needed to get it to explode. However, theorists sometimes overlook little niggling practicalities that make the job harder than they originally imagine. Teller's initial plans for the Super were little more than an atom bomb at one end of a tank of deuterium. The energy from the atom bomb would make the deuterium so hot that the atoms would slam together and stick, fusing and releasing energy. But Teller's deuterium bomb ran into problems from the start. Even deuterium, which is much easier to fuse than ordinary hydrogen, would be hard to ignite. In

1942, mere months after Teller's initial visions of the Super, scientists realized that a "hydrogen bomb" should have tritium, a still heavier version of hydrogen, mixed in with the deuterium if it was to have any chance of exploding. The problem is that tritium doesn't occur much in nature; it has to be manufactured if you want a large quantity of it. And this process required the same resources—and was about as expensive—as manufacturing plutonium.

Even though tritium was scarce, the unlimited power promised by the Super made it worth producing. According to Teller's estimates, it would require a few hundred grams of tritium—a significant, but manageable, amount of the rare substance—to get the Super working. Those estimates were wildly optimistic.

In December 1949, the month before Truman's fateful announcement, the Polish mathematician Stanislaw Ulam, along with his colleague, Cornelius Everett, began extremely tedious calculations to figure out whether, in fact, Teller's Super would work. They worked with pencil and paper, slide rules, and a set of dice.* With each roll of the dice, it became more and more evident that Teller's Super would fail to ignite the tank of deuterium and tritium. Françoise Ulam, Stanislaw's wife, noted how Teller reacted to the ever-worsening news. "Every day Stan would come into the office, look at our computations, and come back with new 'guestimates,' while Teller objected loudly and cajoled every one around into disbelieving the results," she wrote. "What should have been the common examination of difficult problems became an unpleasant confrontation." Ulam wrote that the calculations made Teller "pale with anger." Ulam and Teller already disliked each other, and it was reported that Ulam "took real pleasure" in knocking down Teller's pet project.

By February, it was absolutely clear that the Super wouldn't work. Instead of requiring a few hundred grams of tritium, the Ulam-Everett

---

* They were using a technique that became known as the *Monte Carlo method*; the dice were for generating random numbers that allowed them to get a ballpark solution to a problem much more quickly than an exact calculation would permit.

calculations implied that a Super would need ten times that—a few kilograms. There was no way the United States could manufacture that amount of tritium in a reasonable period of time. Teller's device was impractical.

"Teller was not easily reconciled to our results," wrote Stanislaw Ulam years later. "I learned that the bad news drove him to tears of frustration." Nevertheless, the calculations were solid. It was less than a month after Truman announced the superbomb project—and a month before Truman signed an executive order that officially jump-started the crash program. The Super design was crumbling in front of Teller's eyes.

Ulam and Fermi then put another nail in the coffin. Even with an enormous amount of tritium, they found, a pipe full of the stuff simply wouldn't fuse. If you managed to ignite one end, the reaction would not travel down the pipe. "You can't get cylindrical containers of deuterium to burn because the energy escapes faster than it reproduces itself," explained the Los Alamos physicist Richard Garwin a number of years later. He added that "The classical Super could not work....All the time [on the classical Super] was wasted. There had been miscalculation because Teller was optimistic." After all the commotion and heartache, after the scientific community chose sides in a growing schism, the Super was a fizzle.

The calculations suggested that the enemies of fusion were right all along. "[Teller] was blamed at Los Alamos for leading the Laboratory, and indeed the whole country, into an adventurous program on the basis of calculations which he himself must have known to be very incomplete," wrote Hans Bethe years afterward.

Teller persisted. He had already dreamed up alternate designs for a fusion bomb. One was made out of alternating layers of fissionable heavy atoms and fusionable light atoms. Dubbed the "Alarm Clock" design (because it would wake the world to the prospect of fusion weapons), it had a serious drawback. To make it more and more powerful, designers had to add layer after layer to the bomb. By the time it reached the megaton range, it was too big to deliver. It was not a practical superweapon. It did not promise the unlimited power that Teller was seeking. Neither did his other suggestion. Teller realized that by adding a tiny bit of tritium to the center of an exploding fission warhead, the tritium would fuse and "boost" the yield of the atom bomb. This was a practical idea—and it ultimately did work—but it was just a way of making a slightly better fission weapon. It was far from the thermonuclear fusion superbomb that Teller had promised.

Teller kept up a brave face in public. He tried to recruit scientists to come to Los Alamos to work on fusion weapons. "The holiday is over," he wrote in the *Bulletin of the Atomic Scientists*. "Hydrogen bombs will not produce themselves." However, Teller was fresh out of ideas. Neither the Alarm Clock nor boosted bombs provided the unlimited-power weapons Teller—or Truman—wanted. The crash program was stalling even before it started.

World politics made the situation dire. On June 25, North Korean soldiers marched across the thirty-eighth parallel into South Korea. Seoul fell within days. And within two weeks, General Douglas MacArthur was figuring out how best to use nuclear bombs in the conflict. The battle went back and forth. Then, in November, soon after China entered the war, Truman threatened the use of atomic weapons. MacArthur asked for the discretion to use them on the battlefield. The world seemed on the brink of nuclear war.* The fusion bomb, the weapon that was supposed to restore America's military advantage, was nowhere to be found.

By the autumn of 1950, Teller was desperate. "He proposed a number of complicated schemes to save [the Super], none of which seemed to show much promise," wrote Bethe. "It was evident that he did not know of any solution." The head of Los Alamos, Norris Bradbury, a man whom Teller viewed as an ally of Oppenheimer's, halted design work on the Super until some important tests, scheduled for early 1951, could be run. Teller was furious at the delay. He was at the brink of

---

* The world came very close indeed. After the Soviets and Chinese massed a fresh set of troops on the Korean border in March 1951, the Joint Chiefs of Staff ordered that atomic bombs be used if the Communist troops launched a major new offensive. The bombs were deployed. Truman even signed an order authorizing their use but, luckily, he never sent it.

despair when Bradbury wrote a report summarizing the project for the GAC, Oppenheimer's advisory committee. In Teller's eyes "his report was focused on the Super and was so negative that it seemed an outright attempt to squash the project." Teller and John Archibald Wheeler, a theoretical physicist and fusion hawk, wrote a second report "in a very different tone" to counteract Bradbury's negativity. But there was little way to put a positive spin on the status of the Super. The project was dead in the water. The deuterium wouldn't ignite. Teller was devastated. The unlimited power of fusion was slipping away.

Ironically, it was Ulam, the man who brought Teller to tears, who would lift him out of his despair. Ulam saw a way to build a working fusion weapon. In January 1951, he realized that he could use the stream of particles coming off an atom bomb to compress the hydrogen fuel, making it hot and dense enough to ignite in a fusion reaction.* Instead of a simple bomb with a tank of deuterium, the new hydrogen bomb would have an atom bomb *primary* separated from a deuterium-tritium *secondary*. Particles from the atom bomb—radiation that would ordinarily stream away from the explosion—could be focused onto the secondary to compress, heat, and ignite it. It would be tricky to engineer such a device, but it seemed to overcome the problems that dogged the classical Super. "Edward is full of enthusiasm about these possibilities," Ulam reported to von Neumann. "This is perhaps an indication they will not work." Nevertheless, the enthusiasm was justified. It would mark the end of the dark times for the fusion hawks, and for Teller. By May, Los Alamos would have experimental data to back up the theoretical calculations.

In the Marshall Islands, isolated in deep Pacific waters, a nearly circular atoll of a few dozen islands had been drafted into the Cold

---

* Teller, as often was the case, remembered the situation differently from how his peers did: he says he came up with a solution himself. According to Carson Mark, a weapons designer, "Ulam felt that he invented the new approach to the hydrogen bomb. Teller didn't wish to recognize that. He couldn't bring himself to recognize it. He's taken occasion, almost every occasion he could, not every one, to deny that Ulam contributed anything."

War effort. Since 1948, the United States had used the Eniwetok atoll—some of whose islands were inhabited—for testing nuclear weapons. In April 1951, a new series of tests began. They were code-named Greenhouse.

Greenhouse consisted of four explosions. The first two, Dog and Easy, tested two of the compact fission weapon designs that Los Alamos was furiously generating to keep the United States one step ahead of the Russians. The third and fourth, George and Item, were entirely different. They were the world's first fusion devices.

Greenhouse George was a curious gadget. It wasn't a design that could ever be dropped on an enemy. It was an enormous cylindrical device with a hole in the center. As the device imploded, radiation would stream out of the hole and strike a small target filled with a few grams of deuterium and tritium. It was a science experiment, not a practical weapon, something with which to study the process of fusion rather than to drop on a city. After all, scientists had never achieved fusion before; Greenhouse George, if it worked, would allow them to see it up close for the first time.

It worked. On May 9, Teller, slathered in suntan lotion, watched as a mushroom cloud boiled obscenely into the sky. It was a doozy of an explosion: at 225 kilotons it was a record breaker, an order of magnitude bigger than the bombs that leveled Hiroshima and Nagasaki. George's fission bomb probably generated about 200 kilotons' worth of energy. The remaining 25 kilotons came from the tiny capsule of deuterium and tritium. Scientists had finally unleashed the energy of the sun. Fusing a few grams of hydrogen released the same amount of energy as the fission of many kilograms of plutonium or uranium. And Greenhouse Item, the first test of a fission weapon "boosted" by a little dollop of deuterium and tritium at its center, was also a success. Teller's dream of a weapon of unlimited power was back on track. "Eniwetok would not be large enough for the next one," he gloated.

The next month, Teller and some colleagues met in Princeton to report on their success to Oppenheimer's GAC. The resistance to the fusion bomb—moral and political—crumbled under the evidence provided by the Greenhouse tests. Even Teller admitted that

Oppenheimer was enthusiastic about proceeding. "I expected that the General Advisory Committee, and particularly Dr. Oppenheimer, would further oppose the development [of the hydrogen bomb]," Teller would later testify. But after hearing about the Teller-Ulam design and the results of the Greenhouse tests, "Dr. Oppenheimer warmly supported this new approach." The GAC endorsed a full-scale test of a superbomb.* Nevertheless, Teller saw opposition to the hydrogen bomb everywhere, and Oppenheimer—Teller's Moriarty—was almost certainly behind it all.

Teller's return to Los Alamos should have been triumphant after the success of the Greenhouse tests. Norris Bradbury took the fusion project off its hold and formally started a thermonuclear weapons research program at Los Alamos. The project for the Super was officially back on track. But even this victory contained a defeat. Once again, Teller was passed over—he wasn't appointed the head of the new program. Instead, Bradbury appointed Marshall Holloway, a theoretician. Teller considered Holloway a member of Oppenheimer's clique who "had created difficulties in connection with the hydrogen bomb at every turn." It was a slap in the face, and Teller was horrified to see Holloway and Bradbury dragging their feet on the date of the full-scale test. Teller wanted to see it happen in July 1952; Bradbury and Holloway thought a summer date was too optimistic and scheduled it for later in the year. A week after Holloway's accession, Teller quit. On November 1, 1951, he stormed out of Los Alamos. He hoped to take his fusion program with him.

For more than a year, Teller, along with some of his hawkish allies, had been lobbying to create a second laboratory—one dedicated to thermonuclear fusion. Oppenheimer's GAC consistently opposed the proposal, fearing that it would split the pool of talented physicists

---

* It's not entirely clear why Oppenheimer and others who had expressed such deep moral qualms about the hydrogen bomb in 1949 reversed their position so dramatically in 1951. Oppenheimer said that the idea was so "technically sweet" that the United States had to go ahead and try it and then, later, argue about what to do with it.

rather than keep them concentrated in one place. However, Teller's behind-the-scenes lobbying had yielded some powerful allies, including high air force brass such as James Doolittle, who led the heroic first air raid on Tokyo in 1942. Rather than lose control of hydrogen bomb research to the military, the Atomic Energy Commission began to capitulate. Oppenheimer and the GAC still opposed the new laboratory, but Oppenheimer's influence was waning. New members of the GAC were more hawkish than the ones they had replaced. Worse yet, Oppenheimer's political enemies—Teller, Luis Alvarez, air force scientist David Griggs, AEC director of research Kenneth Pitzer, and GAC member Willard Libby—had been chatting with the FBI about Oppenheimer. Pitzer went so far as to publicly question Oppenheimer's loyalty.

Oppenheimer's critics had taken their toll. In June 1952, he stepped down from the GAC. The following month the AEC green-lighted the second laboratory, to be based at Livermore in California (and is today the Lawrence Livermore National Laboratory). Ironically, Teller was slighted yet again; he was surprised to hear that he wasn't the director. After some arguing he signed up. "I am leaving the appeasers to join the fascists," Teller reportedly joked.

Teller had won. Los Alamos now had a rival, and Teller had a facility that was free from the influence of Oppenheimer's cronies, the Soviet appeasers, the Communist sympathizers. However, it was Los Alamos that would initiate the age of fusion.

At 7:14:59 AM on November 1, 1952, roughly a half second ahead of schedule, the island of Elugelab suddenly disappeared. A compact eighty-ton device, nicknamed "the sausage," unleashed the power of the sun upon the Earth for a few moments. The fusion reaction from this device—the first hydrogen bomb—vaporized Elugelab. All that remained was a cloud of dust and fire that stretched twenty miles into the stratosphere.

The nuclear test, known as Ivy Mike, was the first test of a thermonuclear weapon. The Ulam-Teller design had paid off. The energy it produced was an astonishing ten *mega*tons, fifty times bigger than the Greenhouse George shot and about the size of seven hundred Hiroshima bombs. Eniwetok atoll was now missing an island; Elugelab had evaporated under the cloud of fusing hydrogen, leaving behind a crater that could swallow fourteen buildings the size of the Pentagon.

Teller, the prime architect of the cataclysm, was half a world away. Having left Los Alamos, he was in a darkened room at Berkeley where a seismograph recorded the trembling of the Earth with a tiny beam of light. When that dot of light danced wildly, Teller knew he had succeeded. Ivy Mike had worked. Teller had created a weapon of virtually unlimited power. It was as if the United States had been handed the sword of Michael, the ultimate weapon.

———

It had taken too long. The Russians were already hot on the fusion trail. Shortly after World War II, all across the Soviet Union, mysterious secret cities began sprouting up. Among them: Arzamas-16, near Novgorod; Semipalatinsk-21 in Kazakhstan; Chelyabinsk-70 in the Ural Mountains. After decades of speculation and spying, we now know that these were the cities devoted to designing, testing, and building nuclear weapons. And in the early 1950s, the Soviets were progressing rapidly. Just weeks before Teller left Los Alamos, the third Russian test, Joe-3, yielded forty-two kilotons. Two years later, in August 1953, Joe-4 yielded more than four hundred kilotons. Again, American scientists were shocked. This bomb was more powerful than standard fission weapons—it was clearly a fusion device.

Russian scientists had come up with a design similar to the Alarm Clock idea, the very one that Teller had rejected as a dead end. It was still a dead end; there was no way the Russians could use the design to create a practical weapon in the megaton range. And the fact that they relied on this design shows they hadn't yet made the Ulam-Teller breakthrough. Nevertheless, the Russian alarm clock was waking up the American public to the likelihood that the Soviet Union would have a full-fledged hydrogen bomb in a few years. The American advantage, once again, was dissolving faster than expected.

Teller knew just whom to blame. So did Senator Joseph McCarthy. In April 1954, the senator accused Oppenheimer of deliberately delaying the H-bomb by eighteen months. After years of maneuvering—and after Lewis Strauss, a Teller ally, became the head of the Atomic Energy Commission—Oppenheimer's enemies finally had enough power to break him. The AEC began formal hearings to strip Oppenheimer of his security clearance. The charges against him: various associations with Communists, lying to the FBI about Communist meetings, and strong opposition to the development of the hydrogen bomb in 1949. Oppenheimer was being punished, in part, for not jumping on the fusion bandwagon.

The Communist associations would probably have been enough to sink Oppenheimer.* Nonetheless, Teller and his allies hammered the hapless physicist for dragging his feet about the Super project. Mercilessly. Ernest Lawrence, Luis Alvarez, and Kenneth Pitzer expressed their doubts, in testimony or in affidavits, about Oppenheimer's resistance to building a fusion superbomb. Teller testified, too, and he seemed to relish twisting the knife. "It is my belief that if at the end of the war some people like Dr. Oppenheimer would have lent moral support, not even their own work—just moral support—to work on the thermonuclear gadget…I think we would have had the bomb in 1947." When asked what it would mean to atomic science if Oppenheimer was to "go fishing for the rest of his life," Teller said that Oppenheimer's post–Los Alamos work was simply not helpful to the United States. Scientists sympathetic to Oppenheimer would never forgive Teller for his testimony. Teller likened the reception he got from his fellow physicists to his exile from Europe. He wrote: "In my new land, everything had been unfamiliar except for the community of theoretical physicists.…Now, at forty-seven, I was again forced into exile."

The outcome of the Oppenheimer hearing was almost preordained. The panel stopped short of branding Oppenheimer disloyal, but it revoked his security clearance, stating, among other things, "We believe

---

* Oppenheimer's history was troublesome, especially an incident in 1943, in which, ironically, he alerted authorities to a possible security risk. Oppenheimer told a military officer that a certain person was worth keeping an eye on (and he was), but he lied about the details of how he knew this (through a friend who was a member of the Communist Party). Oppenheimer, when confronted with the lie,

that, had Dr. Oppenheimer given his enthusiastic support to the [Super] program, a concerted effort would have been initiated at an earlier date." Furthermore, the panel found his opposition to the hydrogen bomb "disturbing." Oppenheimer was to blame for the slow progress in building the hydrogen bomb.

The scapegoat had been cast out. Oppenheimer's nuclear career was over.

The bitterness on both sides of the debate would last for decades, but the hawks had won. The weapons project at Los Alamos (and at Livermore) steamed ahead, buoyed by the urgency of keeping out in front of the Soviets. Los Alamos would quickly turn Ivy Mike into a deployable bomb. By 1954, the Castle Romeo test detonated a practical weapon (eleven megatons strong) designed to provide "emergency capability" to U.S. nuclear forces. And there was, in fact, an emergency brewing.

Dwight D. Eisenhower became president in 1953, and like Truman, he threatened to use nuclear weapons against China. In May 1953, American diplomats made veiled but clear nuclear threats that seem to have helped end the Korean War. Even after that conflict was essentially settled, the nuclear saber rattling against China continued. As the United States was drawn into the China-Taiwan standoff, Eisenhower contemplated the use of nuclear weapons. He considered them similar to any other munition, and in March 1955, at the direction of the president, Secretary of State John Foster Dulles announced that nuclear bombs were "interchangeable with the conventional weapons" used by U.S. forces. Dulles also lamented, in a meeting two days later, that a lot of public relations still had to be done with the American people if the nation was to use nuclear bombs within the "next month or two." Luckily, the crisis ended without a nuclear exchange.

Even as the United States used its fusion weapons to try to black-

---

admitted to it in front of the panel: "Isn't it a fair statement today, Dr. Oppenheimer, that according to your testimony now you told not one lie to Colonel Pash, but a whole fabrication and tissue of lies?" asked the AEC attorney. "Right," answered Oppenheimer.

mail China and assert its nuclear primacy, its advantage was slipping away once again. On November 22, 1955, the Soviet Union tested their own Mike: a 1.6-megaton hydrogen bomb. It, too, was a two-stage device. The Soviets had also unshackled the fury of the sun upon the inhabitants of the Earth.

CHAPTER 2

THE VALLEY OF IRON

> ... materials dark and crude,
> Of spirituous and fiery spume, till, touched
> With Heaven's ray, and tempered, they shoot forth
> So beauteous, opening to the ambient light?
> These in their dark nativity the Deep
> Shall yield us, pregnant with infernal flame
>
> —JOHN MILTON, *PARADISE LOST*, VI, 478–83

Nearly a century before the quest for superweapons split the scientific community, fusion was at the center of another debate; physicists just didn't realize it at the time. Decades before fusion was discovered, it was at the heart of an argument between physics and biology, between those who studied the fundamental laws that govern the universe and those who observed the processes of life on Earth. It was a battle between two of the leading scientific lights of the day: William Thomson (also known as Lord Kelvin), and Charles Darwin.

In 1862, Thomson, one of the brightest—and most famous—physicists in Britain, was absolutely certain: Darwin was wrong. The theory of evolution could not possibly be correct. He had ironclad proof.

According to Thomson's calculations, it was not possible for species to change form over millions and millions of years because the sun could not have been around that long. It was a physical fact, Thomson thought, and it would destroy the biologist's pretty theory once and for all.

Thomson's argument was far more ambitious than a mere attack on Darwin's proposal. In fact, the physicist was trying to answer some of the biggest scientific questions of the day. Astronomers were just beginning to understand what the sun was made of, and they were suddenly faced with a new set of questions that once seemed unanswerable. Where did the sun come from? How old was it, really? Where did it get its energy?

When Thomson estimated the sun's age, he calculated that it could only be a few tens of millions of years old, far fewer years than Darwin's natural selection would take to generate the amazing diversity on Earth. It was a major puzzle; two branches of science were giving mutually contradictory answers. It would take decades before scientists uncovered the truth. The secret was fusion. Only when physicists could understand fusion could they understand the nature of the sun, much less create one for their own use.

———

Thomson's calculations were extremely bad news for Darwin. The physicist argued that there was a fundamental problem with the theory of evolution, a problem that seemed to contradict a law of thermodynamics. If true, it would devastate Darwin's theory. The laws of thermodynamics are among the most fundamental and sacrosanct laws of physics, and they brook no contradiction.

The field of thermodynamics studies the relationships among heat, work, and energy. Its first law has to do with energy: energy cannot be created out of nothing. Energy can be transferred from place to place. It can change form. For example, the energy of water spinning a wheel can light a lightbulb: the energy of motion is converted into electric energy then into light energy. However, energy always has to come from somewhere;

it can't be created or destroyed. Nature has a fixed amount of energy, and it is impossible to make more. This law is central to physics, yet Thomson believed Darwin's theory fell afoul of it.

He based his argument on the energy flowing from the sun. When you go out on a bright summer day, you feel the warmth of the sun on your skin. The sun, shining continuously in the sky, transfers some energy to you in the form of light. As you soak up the rays, that energy warms your skin. You are absorbing the sun's energy. Since energy cannot be created or destroyed, that solar energy has to come from somewhere; the sun can't just create energy out of nothing. The sun must be using up some sort of energy reserves. As our star shines, it constantly emits more than 10^{26} watts of power, roughly equivalent to one hundred billion Ivy Mike bombs exploding every second, every day, every year. All that energy is radiated into space.

This baffled Thomson. The sun was emitting energy, and things that emit energy tend to cool down over time. So how can the sun stay so hot? Perhaps it was able to replenish its energy reserves by burning fuel—but what sort of fuel could it be using? It couldn't be burning like a giant charcoal; burning is merely a chemical reaction, and no known chemical reaction could release the sorts of energy that the sun was radiating. No fire could generate that much heat. Could the sun be getting energy from another source—by gravity, perhaps? Meteors occasionally strike the sun, adding energy to the solar furnace, but there aren't that many meteors around, and their energy is just a tiny drop in the tsunami of power coming from the sun. Thomson knew of no possible way to generate the quantity of energy that was escaping the sun every second, yet the first law of thermodynamics dictates that the energy has to come from somewhere.

It was clear to the physicist: no mechanism—chemical, gravitational, electrical, or whatever—existed that could generate the amount of energy the sun was emitting every moment. Since energy can't be created out of nothing, the sun must be depleting its energy reserves at an enormous rate. And that meant that the sun must be slowly getting colder and colder. Unable to replenish its reserves through any means known to man, the sun, presumed to be a gigantic ball of hot liquid, must slowly be cooling down.

Once Thomson reached that conclusion, he wondered: If the sun is merely a huge molten sphere of liquid, where did it get its energy in the first place? The only answer he could think of was the energy due to gravitation. Imagine that the sun came from an enormous cloud of tiny rocks. Those rocks are attracted to one another by the force of gravity. Under their mutual attraction, they begin falling toward one another. As they fall inward toward the center of the cloud, they move ever faster. The cloud of rocks begins to collapse. The individual rocks speed inward quicker and quicker, because their gravitational energy is being converted into kinetic energy—the energy of their motion. As the fast-moving rocks stream toward the center of the cloud and collide, their kinetic energy gets converted into heat energy: the cloud heats up. Eventually, it gets so hot it glows.

Thomson calculated how hot such a protosun could have been. Then he calculated how long it would have taken to cool to its present temperature. Not long. The sun wasn't more than a few tens of millions of years old, not long enough for the long, slow process of evolution Darwin proposed.

In fact, Darwin was deeply shaken by the calculations. He considered Thomson's challenge to evolution "one of the gravest" that the theory had to face, and he could do little to counter it other than argue that scientists did not have a perfect understanding of the nature of the universe.

It was an impasse. The laws of physics seemed to say one thing, while the observations of biologists seemed to tell another.

Physicists would have to follow a tortuous path before they could resolve the contradiction—a path that led, first, to understanding the mystery of matter.

———

By the end of the nineteenth century, physicists and chemists had unraveled many of the mysteries of the universe. Isaac Newton had divined the physical laws that govern how objects move and how gravity works. James Clerk Maxwell had figured out the subtle interrelationships between electric and magnetic forces. Thermodynamicists had codified the laws of energy and heat. At the same time, though, scientists did not know much about matter; they had little idea what sort of stuff made up stars and planets and people. That was soon to change, as they came rapidly to the conclusion that matter was composed of tiny building blocks known as atoms.

Atomic theory, in its most primitive form, goes back to the ancient Greeks. In the fifth century BCE, the philosopher Democritus held that all matter was created out of little indivisible particles. These particles, far too tiny to see, were considered to be uncuttable. Democritus's idea was just one of a huge number of competing theories about the universe. Some philosophers argued that everything was made of fire; others thought that objects were made from a mixture of earth, air, fire, and water. Some argued that matter was infinitely divisible; others, like Democritus, argued that there was a limit to how finely you could slice an object. Though we now know that Democritus's idea was closest to the truth, for millennia it had no special status.

More than two thousand years later, a steady march of experimentation and observation led scientists to the conclusion that Democritus was essentially correct: matter is made up of tiny atoms. Chemists had led the way; the work of chemists such as the Briton John Dalton, the Italian Amedeo Avogadro, and the Russian Dmitri Mendeleev began to produce a picture in which all matter consisted of a collection of invisible "elemental" particles. Water, for example, was made up of two particles of hydrogen and one of oxygen; alcohol had two of carbon, six of hydrogen, and one of oxygen.

There was only a handful of known elements, and they each had different properties. For example, the atoms of some elements, such as hydrogen, oxygen, and carbon, were very light. Other elements' atoms, like those of mercury, lead, and uranium, were very heavy. And these particles—these atoms—were fixed in their properties; it was impossible to transmute an atom of hydrogen, say, into an atom of lead.

This picture explained the nature of matter extremely well. Within a century, atomic theory changed the subject of chemistry from a quasi-mystical hodgepodge of contradictory ideas into a real science. Physicists soon joined the chemists in their support of atomic theory; they began to provide evidence for the existence of tiny atomic particles. Theorists like Ludwig Boltzmann realized that you could explain the properties of gases simply by imagining matter as a collection of atoms madly bouncing around. Observers even saw the random motion of atoms indirectly: the jostling of water molecules makes a tiny pollen grain swim erratically about. (Albert Einstein helped explain this phenomenon—Brownian motion—in 1905.) Though a few stubborn holdouts absolutely refused to believe in atomic theory,* by the beginning of the twentieth century the scientific community was convinced. Matter was made of invisible atoms of various kinds: hydrogen atoms, oxygen atoms, carbon atoms, iron atoms, gold atoms, uranium atoms, and a few dozen others. But, as scientists were soon to find out, atoms are not quite as uncuttable as the ancient Greeks thought. Indeed, to figure out why different elements have different properties, physicists had to slice the atom into pieces.

The first piece came off in 1898. The Cambridge physicist J. J. Thomson was studying a mysterious phenomenon known as cathode rays. He used electric and magnetic fields to deflect the rays and came to the correct conclusion that the rays were made up of negatively charged particles that had been stripped away from atoms. These very, very light particles came to be known as electrons.

Since an atom is, on balance, neither positively nor negatively charged, the positive and negative charges in the atom must be equal and opposite; the charges in the atom have to cancel each other out. This means that for every electron in an atom, there has to be something else in the atom that carries the equivalent positive charge. About a decade after the discovery of the electron, the physicist Ernest Rutherford found out where that equal and opposite charge sits. It resides in

---

* In 1910, the famed physicist Ernst Mach wrote, "If belief in the reality of atoms is so crucial, then I renounce the physical way of thinking, I will not be a professional physicist, and I hand back my scientific reputation."

a tiny, but extremely solid, nucleus at the very center of the atom. This nucleus is quite heavy, thousands of times heavier than an electron, so the nucleus of an atom had to be made of stuff very different from electrons. Rutherford soon figured out what that positively charged stuff was: he realized that the positive charge is cloistered inside a heavy particle known as a proton.

For every electron zipping around in the outer regions of the atom, a proton had to be sitting in the nucleus. Since positively charged objects attract negatively charged ones, the nucleus attracts the electrons through electrical forces, in roughly the same way that the sun attracts its planets with gravitational forces. Rutherford took this analogy fairly literally; he imagined the atom to be like a miniature solar system. At the center is a heavy, dense, positively charged nucleus. Quite a distance away, lighter, quick-moving, negatively charged electrons are in "orbit" around it.* In between, there is empty space—lots of it.

When physicists discovered the proton and electron, they sparked a revolution in the scientific understanding of matter. Two subatomic particles suddenly explained the properties of the elements. No longer were atoms of different elements considered to be fundamentally different objects; an atom of gold need not be thought of as a different sort of creature compared with an atom of lead. Gold and lead were essentially the same kind of object: bundles of protons surrounded by bundles of electrons. Gold has properties different from those of lead—and they both have properties different from the other elements—because they have different numbers of protons in their nuclei (and, hence, different numbers of electrons). A hydrogen atom has one proton per nucleus, helium has two. Oxygen has eight; gold, forty-nine; lead, eighty-two; uranium, ninety-two. In each case, the number of protons in a nucleus—known as an atom's atomic number—determines how the atom behaves chemically. It tells you which atoms it will react with and which it won't; it tells you whether a collection of atoms is likely to be a gas or a metal,

---

* In truth, the analogy is terribly flawed, and electrons don't really "orbit" a nucleus. To explain the behavior of electrons in an atom, you need to get into quantum theory, but this level of subtlety isn't necessary to understand fusion.

whether it will burn in oxygen or explode in water or refuse to react with anything at all. This theory was a tremendous success for science. The uncuttable atom had been dissected into its component parts. But one piece was still missing.

The discovery of the electron had come from Thomson's investigations into cathode rays. Cathode rays come from a fairly simple piece of laboratory equipment: put a couple of pieces of metal in a vacuum tube, hook them to a battery, and *radiation* streams from one end to the other.

The concept of radiation was a new phenomenon at the turn of the twentieth century. Scientists knew little about it, but they were beginning to detect it everywhere. Marie Curie's radium emitted a substance—particles or rays or something as yet unknown—that carried energy; something was fogging a photographic plate. That was one kind of radiation. The German scientist William Roentgen discovered another kind in 1895. When he sent electrical current through an evacuated tube, he noticed it would generate invisible rays that could make fluorescent screens glow. Like the rays coming from radioactive elements like radium and uranium, Roentgen's *x-rays* could expose a photographic plate. X-ray radiation, too, carries energy. (It turned out that x-rays are beams of light so energetic that they pass right through flesh.) Then there were the mysterious rays coming from Thomson's cathode. By the turn of the century, scientists across the world were finding all sorts of rays in strange places. The scientific world was going radiation crazy.

We now know that these "radiations" are not all the same thing. Some, like x-rays, are varieties of light. (Gamma rays, too, are light beams even more energetic, and more penetrating, than x-rays.) These high-energy light rays penetrate matter relatively easily. Not all the radiations had this property. Thomson's cathode rays couldn't penetrate very far into an object before being absorbed. Neither could *beta* radiation, another type of emanation that streams from certain kinds of unstable atoms. Alpha radiation, which comes from yet other varieties of unstable atoms, penetrates even less than beta rays. It turns out that cathode rays, beta rays, and alpha rays are all subatomic fragments. Cathode rays and beta rays are both made up of electrons; alpha rays are made up of heavier, positively charged pieces of large atoms.*

Not surprisingly, researchers were so excited about finding new kinds of radiation that some of their discoveries were entirely fictional. In 1903, the French physicist René Blondlot thought he had discovered a new type, which he dubbed the "N-ray." But Blondlot had deceived himself; his desire to believe in N-rays made him ignore the evidence against them. When a skeptical researcher removed a crucial component of the experimental apparatus and the unsuspecting Blondlot continued to observe the N-rays, N-rays were exposed as a fiction. Blondlot was made a laughingstock.

There was one type of radiation, though, that did not fit neatly into the pattern scientists had been seeing. High-energy light, such as x-rays or gamma rays, penetrates matter easily; its beams are hard to block. Fragments of atoms—charged particles like protons and electrons and alpha particles—tend not to penetrate matter much at all. Because of their charge, they get tangled in the electrons and protons in a given hunk of matter and quickly slow to a halt. But a new type of radiation, discovered in the 1930s, seemed like a weird cross between light and atom fragment. Scientists generated this bizarre radiation by shining a beam of alpha particles upon certain kinds of atoms (such as beryllium atoms). This new kind of radiation did not have an electric charge: it was unaffected by electric or magnetic fields. It penetrated matter as readily as gamma rays did, but it did not behave as a light beam should. It behaved like a heavy particle: it would hit a block of paraffin and knock protons out; mere light couldn't do that so easily. In 1932, the British physicist James Chadwick concluded, correctly, that this new type of radiation consisted of particles almost identical to protons but for one major difference: they had no electric charge. Chadwick won the Nobel Prize for his discovery: the neutron.

The neutron is just a tiny fraction of a percent heavier than a

---

* Technically, these pieces are helium-4 nuclei: two protons and two neutrons all bound together in a tight bundle.

proton, so it has quite a bit of oomph. But because it is electrically neutral, it doesn't "feel" the electrical charges of the electrons and protons in a material. It is only affected by an atom when it slams directly into the nucleus. However, since atoms are mostly empty space and atomic nuclei are very small, a neutron can zoom straight through a chunk of matter without ever encountering something that deflects it. Neutrons penetrate matter extremely well, going through lead bricks almost as if they didn't exist. But when a neutron does, by chance, hit an atomic nucleus, it packs a punch. A light atom (such as hydrogen) might be kicked out of the substance altogether. A heavy atom (such as uranium) might shiver and break apart when struck with the right amount of force. (As described in chapter 1, neutrons doing just this is what causes the chain reaction at the heart of the atom bomb.)

The discovery of the neutron also solved a puzzle that was beginning to vex physicists. When chemists and physicists used their newfound knowledge of protons and electrons to understand the nature of the elements, they were surprised by a strange inconsistency. They discovered that an atom of a given element did not have a fixed weight. For example, in 1932 scientists found that hydrogen came in several flavors. There was ordinary hydrogen—which was thought to be made up of one proton (and one very light electron, whose weight is negligible). Then there was a heavier form of hydrogen that weighed twice as much. They called it deuterium. Soon, they realized there was yet another version that weighed almost exactly three times as much as hydrogen: tritium. All three of these varieties had the same chemistry as hydrogen, but they all had different weights. (And tritium, as it turned out, was radioactive.) Until the neutron was discovered, nothing could explain why a single element could have multiple weights.

Once Chadwick discovered the neutron, though, the answer to the puzzle was obvious. Scientists already knew that the number of protons determined the chemical properties of an atom; hydrogen, deuterium, and tritium each had a single proton in the nucleus, so they were almost identical, chemically speaking. But neutrons can also sit in an atom's nucleus. Because neutrons don't have a charge (and don't attract extra electrons), they don't affect an atom's chemical behavior; an extra neutron doesn't turn hydrogen into a different element. But an extra neutron makes that hydrogen weigh more than before.

Ordinary hydrogen's nucleus is simply one proton. It weighs as much as one proton, so it is known as hydrogen-1, or ^{1}H. Deuterium's nucleus, too, has one proton. But it also has a neutron that weighs roughly the same as the proton; the mass of the nucleus (hence, the mass of the atom) is doubled. Deuterium is thus known as hydrogen-2, ^{2}H. Tritium has a single proton in its nucleus, but in addition it has two neutrons, making it three times as heavy as ordinary hydrogen. Tritium is therefore designated hydrogen-3, ^{3}H. All these atoms are considered to be varieties, or *isotopes*, of hydrogen. In a chemical reaction, all three behave more or less the same way. But they have slightly different physical properties by virtue of their nuclei's different weights.

Scientists were thrilled when they discovered the neutron because it gave them a complete model to explain an atom's chemical behavior. Just figure out how many protons and neutrons are in a given atom and you can predict its properties extremely well.

Despite the spectacular success of atomic theory, scientists, in some sense, were astonished that atoms could exist at all. Nuclei are finicky things, and it is amazing that any of them are stable. By rights, they should fly apart instantly. They are filled with positively charged protons, and positively charged things repel one another. If the protons in a nucleus were to obey their electrical urges, they would flee each other's presence, and the nucleus would explode in all different directions. But something forces the protons to stay put and in close proximity to one another. A very strong force—stronger than gravity, stronger than electromagnetism—glues nuclei together, trapping protons inside. In a great burst of creativity, scientists dubbed this strong force...the *strong force*. This force holds the secret to nuclear fusion.

The strong force is powerful enough to overcome the natural repulsion that protons have for other protons. However, it can do so only under a fairly narrow range of conditions. If there is the right balance of particles in the nucleus—the correct number of protons and neutrons—the strong force keeps the nucleus stable (or nearly so), preventing the nucleus from exploding. If there are too many neutrons or too few, the nucleus will be unstable. An unstable atom will destroy itself somehow, changing the balance of particles in its nucleus until the nucleus reaches a more stable state. A nucleus can break apart, spit a particle out, or swallow one to get closer to an ideal, stable balance of protons and neutrons.

For example, hydrogen (one proton) and deuterium (one proton and one neutron) are stable. Left to their own devices, they would not change at all. But add a second neutron to the mix, making tritium, and the atom has too many neutrons for comfort. It is no longer stable. Eventually, a tritium atom will, spontaneously, transmute one of its neutrons into a proton (and spit out an electron in the process). The substance left behind is no longer tritium; it has become helium-3, a stable if rare isotope of helium that has two protons and one neutron. (Most helium is helium-4, which has two protons and two neutrons.) It takes an average of twelve years or so for any given tritium atom to undergo this *decay* process, but over time, if you have a jar full of hydrogen-3, you will find that it slowly transforms itself into helium-3.

This transformation process releases energy, because the helium-3 does not weigh exactly the same as the tritium did. The neutron that disintegrated weighed more than the proton and the electron that it turned into.* There is mass missing: it was converted into energy, just as $E = mc^2$ says. As the unstable tritium changed itself into the stable helium-3, it lost a little bit of mass and released a bunch of energy.

This is an example of a general rule. When a nucleus converts itself from a less-stable variety to a more-stable one, it releases a little bit of energy because some of its mass disappears. And nuclei always "want" to become more stable, just as a ball perched on a hill "wants" to roll

---

* Technically, a third particle known as an antineutrino is also created.

down to the bottom. In the process of getting more stable, an atom releases energy, just as a ball rolling down a hill picks up more and more speed as it goes.

Marie Curie was seeing this process with radium. Radium-226 is a heavy atom with 88 protons and 138 neutrons. It is almost stable...but not quite. On average, after 1,600 years, a radium-226 nucleus spits out an alpha particle (a helium-4 nucleus: two protons and two neutrons), leaving behind 86 protons and 136 neutrons—radon-222—and releasing a bunch of energy. This energy heats the hunk of radium, and it is why Curie observed that chilled radium would warm itself. It is also why a hunk of radium emits radon and helium. It so happens that radon, itself, is unstable; it decays into thorium, releasing energy, which, in turn, decays into another species and another and another, emitting energy at each step. Like a ball rolling down a bumpy hill, it keeps rolling and rolling until it reaches a stable place to rest: in this case, lead-206, which is much more stable than radium-226. The ball has rolled a long way down the hill. But it didn't roll *all* the way down. There are, in fact, nuclei more stable than lead-206. At the very bottom of the valley are the most stable atoms of them all: the iron group.

Iron-56 (26 protons, 30 neutrons), nickel-62 (28 protons, 34 neutrons), and a few other nearby iron and nickel isotopes are the ne plus ultra of the nucleus world. They are the most stable elements of them all. They are at the very bottom of the valley. All other atoms "want" to be iron, just as a ball anywhere on the slope of a hill "wants" to be at the very bottom.

The landscape of nuclei is very much like a valley with a steep hill on one side and a shallow hill on the other (see the graph on page 47). At the very bottom of the valley is iron. Heavy nuclei, like radium and uranium, have more protons and neutrons than iron—they are high up on the shallow hill. To roll down to iron, they have to get lighter, shedding those extra protons and neutrons. Sometimes they do it in small steps, like radium does. Sometimes they do it violently, by breaking into two or more parts: fission. (Indeed, fission is little more than a way for heavy atoms to roll quickly down the shallow hill, losing mass and releasing energy in the process.) Light elements, on

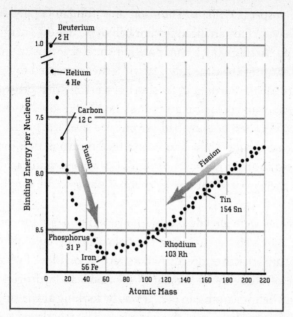

CURVE OF BINDING ENERGY: Iron is at the bottom of the valley. Light elements on the left and heavy elements on the right release energy when they move down the valley of iron by fusing or fissioning.

the other hand, are on the steep hill. They have to get bigger if they want to get into the valley of iron. This is what fusion is all about. Two light nuclei, if they slam together, can stick to one another to create a larger nucleus.

Fusion is a way for light atoms to roll down the steep hill toward iron. Since the fusion hill is much steeper than the fission one, a fusion reaction yields much more energy than an equivalent fission reaction. Fusion and fission are two sides of the same coin, but as Teller well knew, fusion is more powerful than fission.

In the 1930s, fusion was soon to solve the puzzle that had so vexed Darwin and Kelvin, and it would answer a question that had bothered humans for millennia: Why does the sun shine? Hans Bethe, then a physicist at Cornell University, would uncover the answer.

In retrospect, William Thomson was fundamentally correct. If the sun were, in fact, an incandescent ball of liquid, the energy released by infalling matter would only power it for a few tens of millions of years, far short of the time that Darwin's theory needed to explain the diversity of life on Earth (and far short of the time that other scientists needed to explain geological processes). But Thomson's work preceded $E = mc^2$ by decades; nobody had yet puzzled over the nature of radioactivity or understood how fission and fusion turn matter into energy. Fusion held the solution to Thomson's puzzle and vindicated Darwin. Fusion is the source of the energy that has powered the sun for billions of years.

The clues were already old by the time Bethe set to work on the problem. By carefully analyzing the colors of the light that streams from the sun, scientists already had a pretty good idea of what the sun was made of. Roughly 90 percent of its atoms are hydrogen. About 9 percent are helium atoms; in fact, it was by looking at the sun that scientists discovered helium in the first place. The remaining 1 percent is mostly carbon, nitrogen, oxygen, neon, and a tiny smattering of heavier elements, but almost all of these are lighter than iron. The sun bears all the hallmarks of being powered by fusion. Bethe figured out precisely how that power is generated.

A star begins its life as a cloud of gas: mostly hydrogen and a little bit of helium. Because atoms have mass, they attract each other gravitationally, and because of this mutual attraction, the cloud begins to collapse under its own gravity. As gravity compresses the cloud, the cloud heats up.

If gravity were the only force at play, the cloud would simply get smaller and smaller and eventually collapse into a tiny, massive point. But that is not what happens. As the gas cloud gets denser, atoms of hydrogen bump into each other more and more frequently. The collision rate increases dramatically. And as the cloud heats up, its atoms have more energy and collide more violently. The hydrogen atoms jostle each other harder and harder.

Ordinarily, nuclei try to escape from one another. They are positively charged, so they find other nuclei repulsive. When two atoms "collide," they don't usually come into physical contact. Once they get within close range, the repulsive forces send them zooming in opposite directions before they actually touch—something like what happens when you try to make two powerful magnets touch each other despite their mutual repulsion. But if the nuclei are moving fast enough—if both atoms are hot enough—then even the mutual repulsion is not enough to keep the nuclei from hitting each other. The two nuclei slam together with great force. This is where fusion begins, and how a sun sparks to life.

Hans Bethe realized that with all these hydrogen nuclei constantly slamming into one another, two hydrogen nuclei—two protons—might smash together at the same time that one spits out a set of particles, turning itself from a proton into a neutron. They fuse, creating deuterium and releasing energy in the process. The deuterium slams into another proton, making helium-3, and again releasing energy. And when two helium-3 atoms collide with each other, they fuse, making helium-4 and releasing two protons and yet more energy. This process, known as the proton-proton chain, turns four hydrogen atoms into helium-4 and lots of energy. Bethe figured out that this was one way

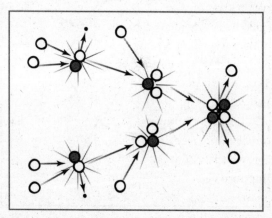

FUSION REACTIONS IN THE SUN: Colliding protons release an electron and make deuterium, then deuterium and protons make helium-3, and finally helium-3s make helium-4, producing a lot of energy in the process.

our sun generates power: by turning hydrogen into helium. He also realized that other processes are going on as well; for example, the trace amounts of carbon, nitrogen, and oxygen are involved in a cycle that has the same outcome as the proton-proton chain: this process takes four hydrogens and turns them into helium-4. Once a cloud of hydrogen gets hot enough and dense enough, it turns into a machine that converts hydrogen to helium, releasing energy. That is how the nuclear furnace at the heart of a star works.

The fusion energy released in the guts of the sun makes it shine. But it also threatens to blow the sun apart. The energy heats the hydrogen gas, making the nuclei slam together harder and harder, and the reaction speeds up, pouring more energy into the cloud. This would appear to lead to a runaway reaction; the furnace should run hotter and hotter and eventually get so energetic that the cloud explodes violently in all directions. However, it turns out that the hotter a cloud of gas

EQUILIBRIUM IN A STAR: Two forces compete for dominance in every star; gravity tries to crush the star to a point, while the fusion explosion tries to blow the star apart. In a normal star, these two forces are in balance.

is, the more it expands. So when the fusion engine runs hot, the star expands slightly. It becomes slightly less dense and the atoms slam into each other less and less often. The fusion engine slows, and the star cools. Gravity takes over once more, compressing the star, heating it up, and making the fusion energy run hot again. This means that a star is in a delicate equilibrium, caught between the force of gravity and the energy of fusion. The force of gravity tries to collapse the star while the energy of fusion tries to blow it apart.

When that delicate equilibrium fails, the star dies. A fusion engine, no matter how well-balanced, can only run for as long as it has fuel. As a star gets older, its hydrogen supply begins to run out; the hydrogen fusion cycles sputter to a halt. A large star then turns to other light elements to keep itself from collapsing. It begins to fuse helium, turning it into yet heavier elements, such as carbon and oxygen. As the helium runs out, the star fuses heavier and heavier fuels: carbon, oxygen, silicon, sulfur. The fusion engine is rolling further and further down the fusion hill. Soon it hits bottom. The valley of iron.

Fusion gets its energy by making light elements roll down the hill toward iron. Fission gets its energy by making heavy elements roll down the hill toward iron. Iron, already at the bottom of the hill, can't yield energy through fusion or fission. It is the dead ashes of a fusion furnace, utterly unable to yield more energy. When a star runs out of other fuels, its iron cannot burn in its fusion furnace. The fusion engine has nothing that it can turn into energy, so it shuts off and the star abruptly collapses. Depending on the star's nature, it can die a fiery death: the final collapse ignites one last, violent burn of its remaining fuel, blowing up the star with unimaginable violence. A supernova, as such an explosion is called, is so energetic that a single one will typically outshine all the other stars in its galaxy combined. The star spews its guts into space, contaminating nearby hydrogen clouds in the process of collapsing into new stars. This is what happened before our sun was born; it got seeded with the nuclear ash of a supernova explosion. All the iron on Earth, all the oxygen, all the carbon—almost all the elements heavier than hydrogen and helium—are the remnants of a dead fusion furnace. We are all truly made of star stuff.*

—————

When Hans Bethe solved the riddle of the sun's energy, the idea of a fusion bomb seemed absurd. To ignite a fusion reaction, you need to have a bunch of light atoms that are extremely hot (so they have enough energy to overcome their mutual repulsion and slam into each other) and extremely dense (so they are close enough to one another that they collide frequently). The laws of nature seem to conspire against having those two conditions at the same time: hot things expand, reducing their density. The only reason a star can keep a stable fusion engine going is because it is so massive. It is held together by the intense force of its own gravity, resisting the explosive force of the fusion engine in its belly. A cloud of hydrogen smaller than a star doesn't have the benefit of that gravitational girdle keeping the reaction from puffing out. Even if you are somehow able to start a fusion reaction, it will blow itself up and snuff itself out in moments.

It is extraordinarily hard to get fusion going outside a star. Even the biggest explosion of all—the big bang—couldn't get fusion going for more than a few minutes. In the first seconds after the big bang, all the matter in the universe—lots of fundamental particles, including a whole bunch of protons—was contained in a relatively small, intensely hot space. Protons—hydrogen nuclei—fused to create helium. But the universe was expanding rapidly because of its own explosive energy.

—————

* The fusion furnace accounts for the abundance of light elements. But if fusion can't fuse nuclei to get atoms heavier than iron, where do we get gold and lead and uranium from? It turns out that they are created in the very final moments of a star's life. As the star explodes, the explosion is so hot and so violent that heavy nuclei are colliding with protons, neutrons, and other particles with great force. Sometimes, the particles stick, making the nucleus bigger. This process absorbs energy rather than releasing it—it's like rolling the ball up the shallow side of the hill—but it happens because the explosion is so energetic. It is from the very last moments of a supernova that we get all the elements heavier than those in the iron group.

After about three minutes, the universe had expanded so much that the matter wasn't dense enough to fuse anymore. As hot and dense as the early universe was, it could only sustain fusion for a measly three minutes. After that, there wasn't any fusion going on in the universe, not until it occurred again in the heart of a solar furnace. To get fusion going on Earth, you must create conditions that are as hot as the first few minutes after the big bang. And without the massive gravity of a star, it is nearly impossible to keep those conditions going for very long.

This was the major obstacle to designing the hydrogen bomb. Even with deuterium and tritium (or lithium) as fuel—deuterium and tritium are relatively easy to fuse—it is hard to make the fuel hot enough and dense enough to get the nuclei fusing. And if you can initiate fusion, you have to maintain the high temperatures and high density long enough to generate an appreciable amount of energy from it.

Teller's first design would not have worked. Even with the power of an atomic weapon behind him, he would have found it difficult to get the hydrogen fuel hot enough and dense enough to ignite. As a substance heats up, it radiates energy more rapidly. In fact, the radiation goes up as the fourth power of the temperature: double the temperature of an object and it radiates its energy sixteen times as fast. To ignite fusion, the fuel has to get to tens or hundreds of millions of degrees (depending on the density of the fuel). Yet even then, it will radiate its energy away at a tremendous rate; it is almost as if everything in the universe is trying to cool it down. And if he had been lucky enough to ignite fusion, he would have been unable to keep the reaction from blowing itself apart with its own energy just as it got going.*

The Alarm Clock design was the simplest way to get around these problems. This bomb was to be like a spherical layer cake with alternating layers of heavy, fissioning material and light, fusionable hydrogen isotopes. By imploding the whole thing symmetrically, making sure that nothing squirted out, the hydrogen would get hot and dense

---

* A fusion reaction that isn't properly compressed becomes a big, expensive dud. In weaponeers' terminology, the bomb "fizzles." Livermore's first nuclear tests fizzled, including its first hydrogen bomb test, Castle Koon.

enough to ignite. It would fuse for a tiny fraction of a second before the whole package blew itself apart. Teller dreamed up the concept in 1946 and discarded it as impractical. In Russia, the physicist Andrei Sakharov came up with a very similar design and called it the "sloika" after a Russian layer cake. The sloika was the basis of the Joe-4 test, but the design was eventually abandoned because megaton-size bombs became too large to use as weapons. Within a few years, Sakharov, like Teller and Ulam in the United States, figured out a much cleverer way to ignite a fusion reaction for a short time.

On the surface, it might seem that America's first fusion bomb, Ivy Mike, was little different from Teller's original "bomb at the end of a tank of hydrogen" design. But in fact, it was impossible to ignite a cylinder full of fuel in the way that Teller had hoped; more energy was radiated away by the expanding fireball than the fusion was producing, so the reaction would snuff itself out very quickly. The Teller-Ulam design had some subtle techniques to avoid this problem.

In a bomb like Ivy Mike, the fission bomb that starts the reaction (the primary) is a distance away from the cylindrical vessel containing deuterium and tritium (the secondary). As the fission bomb explodes, it radiates a huge number of x-rays in all directions. These x-rays, being light waves, travel at light speed and move much faster even than the blast wave coming from the fission bomb. As the atom bomb explodes, the x-rays course through a channel left in the casing that houses the primary and secondary. The x-rays then vaporize a plastic shell, turning it into a plasma, a hot soup of nuclei and electrons. This superhot plasma radiates more x-rays, which strike a heavy *pusher* surrounding the fuel, compressing the fuel from the outside. As the fuel cylinder compresses, it heats up, getting denser and denser. The compressing plasma soon ignites a small "spark plug" of fissionable material at the center of the cylinder, generating a second fission explosion that squeezes the fuel from the inside. The deuterium and tritium are caught between a pusher that is pushing inward and the spark plug explosion pushing outward. The fuel compresses even further and, whoomp! The fusion reaction ignites. It's as if there's a tiny hunk of the sun on Earth.

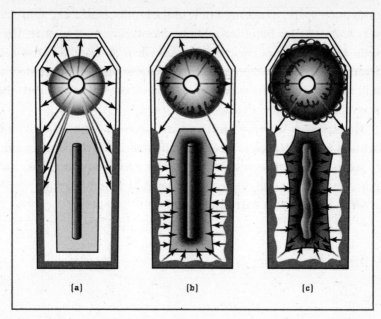

DETONATION OF A HYDROGEN BOMB: (a) A fission bomb explodes at one end of the device, sending x-rays in all directions. (b) The x-rays strike the walls of the container, causing them to evaporate and radiate more x-rays. These x-rays hit the container containing deuterium and tritium, causing it to implode. (c) The compressing deuterium and tritium fuel heats up and ignites a fission "spark plug" at the center of the device, causing it to explode outward. Trapped between the imploding container and the exploding spark plug, the fuel ignites in a fusion reaction.

The reaction only lasts for a fraction of a second before it blows itself apart, but in the process it releases an enormous amount of energy. Ivy Mike was equivalent to ten megatons of TNT, but there was no reason why the whole device could not have been scaled up by adding a third stage…or a fourth. In 1961, the Russians detonated a (roughly) fifty-megaton whopper nicknamed the Tsar Bomba, the most powerful weapon ever built by man.

In theory, there was no end to the power of fusion. But the race to build ever-bigger hydrogen bombs crept to a halt, because it had diminishing returns. As early as 1949, scientists realized that after about 150 megatons, hydrogen bombs simply take a huge column of air and lift it into outer space, punching a hole in the atmosphere about fourteen miles across. Bigger bombs would not do much more than that. They would radiate most of their energy uselessly into space. So after 150 megatons, there was no point in getting bigger, unless you wanted to build a fusion device large enough to destroy the Earth. Not even the most rabid hawks were in favor of that.

Nevertheless, with Ivy Mike and its successors, the fusion bomb scientists had succeeded at creating a tiny star on Earth. For a fraction of a second, scientists were able to get a fusion reaction going. They had figured out how to use that energy for war. It would be much, much harder to harness that energy for peace.

CHAPTER 3

PROJECT PLOWSHARE
AND THE SUNSHINE UNITS

I've felt it myself, the glitter of nuclear weapons. It is irresistible if you come to them as a scientist. To feel it's there in your hands to release this energy that fuels the stars. To let it do your bidding. To perform these miracles, to lift a million tons of rock into the sky. It is something that gives people an illusion of illimitable power and it is in some ways responsible for all our troubles—I would say, this, what you might call "technical arrogance" that overcomes people when they see what they can do with their minds.

—FREEMAN DYSON, IN *THE DAY AFTER TRINITY*

On January 15, 1965, deep inside the Soviet Union, a nuclear rumble shook the earth. The Americans were the first to spot the radioactive cloud as it floated over Japan and toward the Pacific Ocean beyond. The Soviets had detonated a fusion bomb, and the fallout was contaminating the atmosphere over the Japanese mainland. This particular explosion was different from all the other bombs exploded over two decades of nuclear testing. This one had the makings of a major diplomatic incident.

A little more than a year earlier, the United States and the Soviet Union had signed an international agreement to limit the scope of nuclear testing. No longer was it acceptable to detonate a nuclear bomb on the surface of the Earth, in the atmosphere, underwater, or in space. Only underground testing was allowed. Even an underground test was a violation if it caused radioactive debris to float across national borders. The fallout dropping on Japan was clear evidence that the USSR was violating the treaty.

The Americans had few details about the test. At first, their seismic monitors seemed to imply that it was a 150-kiloton bomb that had exploded underground but which had been imperfectly contained, allowing some of the radiation to escape into the atmosphere. Within a few days, American scientists had revised their guess, estimating that the bomb might be as big as a megaton. They also knew roughly where the explosion happened: in Kazakhstan at the Semipalatinsk test site. They knew little else.

On January 21, Secretary of State Dean Rusk demanded an explanation from the Soviet ambassador, Anatoly Dobrynin. Dobrynin responded, "An underground explosion was indeed carried out in the Soviet Union…deep down underground." Furthermore, he insisted, the amount of debris that leaked into the atmosphere was "insignificant." It was a lie.

The explosion, and the resulting fallout, was the beginning of a top-secret Soviet project, Program No. 7. Starting with the mysterious January 1965 explosion and continuing for the next twenty-three years, Program No. 7 would use fusion weapons to dig canals, build underground storage caves, turn on and shut off gas wells, and change the face of the Earth. With the January 1965 explosion, Russian scientists, in a fraction of a second, had carved a major lake and a reservoir, now known as Lake Chagan, out of bedrock.

Program No. 7 was not the only secret government project to harness the power of fusion. An equivalent program was already under way in the United States. A few years earlier, American scientists began work on Project Plowshare and started drawing up plans to use nuclear weapons to create an artificial harbor in Alaska, widen the Panama Canal, and dig a second Suez canal through Israel's Negev desert.

Project Plowshare and Program No. 7 were crude attempts to harness the power of fusion. Researchers quickly reasoned that if humans could learn to control the power of fusion, it could be the biggest boon that mankind has ever seen. We could visit the outer reaches of the solar system and even visit nearby stars. We would never have to worry again about dwindling energy supplies, oil crises, or global warming.

Of course, it wouldn't work out quite that nicely.

In the early 1950s, the world seemed on the brink of nuclear war. Mankind had unleashed a force so great that it could destroy entire cities in a fraction of a second. And though the United States had a small lead over the Russians in developing superbombs, soon both sides would be armed with fusion weapons. If nothing was done, said President Eisenhower in 1953, humanity would have to accept the probability of the end of civilization. At the same time, civil defense films, while trying to calm a jittery nation, whipped up an overwhelming fear of nuclear annihilation. The scientists who had unleashed the power of the sun had placed a great burden on humanity.

But they had hopes that their work would save civilization rather than destroy it. With this unimaginable source of energy, they could usher in a golden era. Even as Bert the Turtle implored children to "duck and cover" upon sighting the inevitable flash from a Russian bomb, other films touted the brilliant future revealed by nuclear power. In the 1952 short *A is for Atom*, a giant glowing golem, arms crossed, represented "the answer to a dream as old as man himself, a giant of limitless power at man's command." And Eisenhower, for all his talk of nuclear annihilation, envisioned an earthly utopia if we put the power of the atom "into the hands of those who will know how to strip its military casing and adapt it to the arts of peace."

The paranoid, anti-Communist Edward Teller was the man who

most desperately tried to bring us to the promised land. He and his allies lobbied for more and more money to figure out how to harness the immense power of fusion. Lewis Strauss, the AEC chairman and Teller backer, promised the world a future where the energy of the atom would power cities, cure diseases, and grow foods. Nuclear power would reshape the planet. God willed it. The Almighty had decided that humans should unlock the power of the atom, and He would keep us from self-annihilation. "A Higher Intelligence decided that man was ready to receive it," Strauss wrote in 1955. "My faith tells me that the Creator did not intend man to evolve through the ages to this stage of civilization only now to devise something that would destroy life on this earth."*

Unfortunately for Teller and the other fusion aficionados, it wasn't easy to use fusion for peace. Fission, not fusion, was the obvious choice for nuclear energy. Ever since Enrico Fermi built a nuclear reactor in the Chicago squash courts, scientists have been able to use uranium to generate power. By controlling the rate of the fission in a reactor, engineers could generate as little as half a watt of power, barely enough to light a Christmas light, or as much as a few hundred million watts of power, enough for a small city. Engineers were drafting plans to build nuclear ships, nuclear submarines, nuclear locomotives, and even nuclear airplanes. But the potential of fission seemed microscopic compared to the unlimited power of fusion, and this is what excited Edward Teller so much. Fusion couldn't just generate energy, it could move mountains. Literally. Teller was going to make it happen. "If your mountain

---

* While fusion hawks like Teller and Strauss led the push to turn weaponry into something to benefit mankind, even the scientists on the other side of the hydrogen bomb divide—those who opposed the development of the Super—pushed to turn nuclear knowledge into a boon for humanity. "I had a hand in formulating and popularizing that hope of peaceful potentials," wrote the former AEC chairman (and Oppenheimer ally) David Lilienthal. "The basic cause, I think was a conviction, and one that I shared fully, and tried to inculcate in others, that somehow or other the discovery that had produced so terrible a weapon simply *had* to have an important peaceful use."

is not in the right place," he once said at a press conference, "drop us a card." He meant it. He was hoping for the chance to show what fusion could do.

In 1956, world politics provided just such an opportunity. In July, the Egyptian government nationalized the Suez Canal, sparking an international crisis. Britain, France, and Israel attacked Egypt, and the situation threatened to spin out of control. Thanks to the intervention of the United Nations, the crisis was resolved, but Western strategists were clearly frightened. The prospect of a crucial waterway in the hands of a nationalist Arab government seemed like a ticking time bomb waiting to explode into a major war. Even though the Suez crisis had been brought under control, the threat of a Suez blockade remained.

Teller and his Livermore colleagues immediately seized upon Suez as an opportunity; they announced that fusion could solve the Egyptian problem. A promising young Livermore scientist, Harold Brown, argued that engineers could use the power of fusion to dig a second canal, eliminating the Suez threat once and for all. Brown—who would later become the secretary of defense in President Jimmy Carter's administration—figured that if a chain of hydrogen bombs, exploding across Israel's Negev desert, cut a second channel from the Mediterranean to the Red Sea, Egypt would no longer have a monopoly. Fusion energy would build a canal in the territory of a Western-friendly power. Teller realized that a new Suez was just the beginning. Fusion weapons could move great volumes of earth, completely reshaping the world's topography to benefit mankind. In February 1957, Livermore hosted a conference to develop the idea of peaceful nuclear explosions and to solicit ideas for nuclear engineering projects.

Of course, many scientists were skeptical of the whole concept; the idea of using hydrogen bombs for peaceful purposes seemed patently absurd. Isidor Rabi, who had called the hydrogen bomb an evil thing under any light, huffed incredulously to Brown, "So you want to beat your old atomic bombs into plowshares?" Rabi's ironic comment harked back to the prophet Isaiah's bright vision of a coming paradise on earth: "they shall beat their swords into plowshares and their spears into pruninghooks: nation shall not lift up sword against nation, neither shall they learn war any more."

Brown—and Teller—turned Rabi's irony into pure optimism, and embraced Isaiah's vision. Project Plowshare was born. Fusion power, even in a vessel as crude as a hydrogen bomb, could make the world a better place. The Livermore scientists quickly set to work figuring out what engineering projects were suitable for nuclear ditch-digging.

The ideas started coming. Build a new Suez. Dig a new Panama Canal. Cut a waterway across Thailand. Excavate a harbor in North Africa or in Alaska. Blow up rapids to make rivers navigable. Cut trenches to help irrigate crops. Straighten the route of the Santa Fe Railroad. Mine coal and rare minerals. Free oil and gas reserves. "We will change the earth's surface to suit us," Teller wrote. Mines and trenches were just the obvious applications. Teller also suggested using hydrogen bombs to change the weather, to melt ice to yield fresh water, and to mass-produce diamonds. (Another unconventional suggestion attributed to him was to close off the Strait of Gibraltar, making the Mediterranean a lake suitable for irrigating crops.) Ted Taylor, a bomb designer, argued that nuclear bombs would be able to drive a rocket into deep space, even to other stars.* Teller even found the idea of bombing the moon incredibly enticing. "One will probably not resist for long the temptation to shoot at the moon…to observe what kind of disturbance it might cause," he wrote.

By 1957, scientists had scads of ideas for peaceful uses of hydrogen bombs. The next step was to figure out whether these grand schemes could possibly work. Could fusion bombs carve canals and harbors, much less turn the Mediterranean into a freshwater lake? They could only find out by running experiments.

---

* Taylor had less peaceful uses in mind, too. The neutrons generated by an exploding fusion bomb buried under the ice could generate oodles of tritium; such a bomb exploded over a blanket of uranium could manufacture all the plutonium that the defense industry could possibly need.

In September 1957, the United States performed an underground weapons test: Plumbbob Rainier. A small nuclear bomb, only 1.7 kilotons, was buried under the surface of the Nevada desert. When the device went off, the earth jumped a few inches and then settled. Scientists later saw that the bomb had vaporized rock to make a one-hundred-foot hole underground. From the Plowshare scientists' point of view, it was a stunning success: nuclear bombs could indeed break up rock and change the landscape, with little release of radiation into the environment. It was time to try to change the Earth.

During the summer of 1958, Edward Teller flew to Alaska to unveil a new "nuclear engineering" project: Project Chariot. Using two large one-megaton bombs and four smaller hundred-kiloton ones, Teller hoped to carve a large harbor on the northwestern coast of Alaska. He pitched the project as an economic boon: the harbor would help Alaskans with fishing and with transporting Alaskan coal by sea. Locals were very skeptical. They had good reason to be.

Despite Teller's slick sales job, the harbor made little economic sense. It would be icebound for most of the year, no substantial fishing was done nearby that would be helped by a harbor, and the coal would have to be transported by rail to the docks—via a railroad that would cost a staggering $100 million to build. Alaskans were wary of Teller's grand scheme for another reason, too. Fallout.

———

Ever since Hiroshima, scientists had known of the deadly aftereffects of nuclear weapons. The atomic bomb had left thousands crippled—burned and blighted by the invisible radiation that streamed from the bomb, harboring cancers and genetic defects that would linger for years after the war had ended.

An exploding nuclear bomb is a veritable treasure trove of radioactive debris: the unfissioned uranium and plutonium from a bomb's primary as well as lighter radioactive atoms left behind by the uranium and plutonium that did fission. A great burst of neutrons also accompanies a large blast; these neutrons strike surrounding atoms—in the atmosphere, in the dirt, in people—with great force. Occasionally these neutrons stick, changing once-stable atoms into radioactive ones. Neutrons can turn ordinary material into a radioactive mess, a phenomenon known as neutron activation. Neutron-activated material, catapulted high into the air, falls to earth downwind of a nuclear explosion, irradiating anyone unfortunate enough to come into contact with this fallout. (Radiation strips electrons from DNA and alters its structure, killing cells and causing cancers.) If a nuclear explosion is powerful enough, it sends radioactive debris so high into the atmosphere that fallout can descend halfway around the globe.

As radioactive as the Nagasaki and Hiroshima bombs were, the multimegaton blasts of fusion weapons were much worse. The world got a taste of their deadly potential in 1954 with the Castle Bravo nuclear accident.

At 6:45 AM on March 1, 1954, the United States detonated a hydrogen bomb; ground zero was a reef in Bikini atoll. The explosion was much bigger than expected—fifteen megatons, the largest explosion yet—roughly equivalent to one thousand Hiroshima-sized bombs. The fireball pulverized the coral reef, sending pieces flying thousands of feet into the air.

By 8:00 AM, "pinhead-sized white and gritty snow" began to shower the American fleet observing the test. This was highly radioactive fallout. The radiation levels on the ships rose rapidly, and the fleet immediately steamed south to escape, but not before more than twenty sailors received radiation burns, and thousands more had been exposed to fallout. Fifteen minutes later, the snow began to fall on a Japanese fishing vessel, the *Daigo Fukuryu Maru*. The whole crew was exposed. (The captain died shortly thereafter, the first person killed by a fusion weapon.)* Within hours, the eastward-drifting cloud dropped fallout

---

* He would not be the last, nor would Castle Bravo be the only fusion "oops." When the Soviets detonated their first Ulam-Teller-type device in 1955, a temperature inversion in the atmosphere reflected the shockwave back to the ground, causing a tremendous amount of damage. A Russian soldier died when his trench collapsed, and in a nearby settlement a two-year-old girl, who had been playing with blocks, was killed when the shockwave smashed the bomb shelter she was in.

on the Rongelap atoll and some other nearby, inhabited islands. The navy evacuated more than six hundred people, many of whom developed "raw, weeping lesions" from the radiation.

It was a public relations nightmare. AEC chairman Lewis Strauss tried to reassure the public that the island natives were "well and happy," but it was hard to hide the truth, and the photographs of burned islanders, from the press. The newspapers had lurid details; they even told of how the ship's cargo of radioactive fish was put up for sale on the Japanese market. (A *New York Times* subhead, "Radioactive Fish Sought In Japan," seemed like something from a B movie. It was hardly good press for American nuclear scientists.)*

The Castle Bravo accident marked a turning point in the perception of nuclear tests. Every time such a test weapon exploded, it spewed radioactive ash into the atmosphere, and scientists noticed that the world was becoming increasingly radioactive as a result. As tests continued, the problem got worse. Scientists were particularly concerned about a radioactive isotope of the metal strontium: strontium-90. Produced by fission in an atomic or hydrogen bomb, strontium-90 is metabolized in a way similar to calcium. It is readily taken up by the body, especially a child's body, and is deposited in bones, teeth, and mother's milk. Once it is inside the body, it destroys from within. (Nuclear scientists measured strontium-90 dosages in "sunshine units," but the cheery name didn't reassure anybody.) And observers were detecting more and more strontium-90 worldwide.

By the mid-1950s, scientists such as Albert Schweitzer and Linus Pauling were raising the alarm. "Each nuclear bomb test spreads an added burden of radioactive elements over every part of the world," read a Pauling-drafted petition from 1957. "Each added amount of radiation causes damage to the health of human beings all over the world and causes damage to the pool of human germ plasm such as to lead to an increase in the number of seriously defective children that

---

* Despite the bad press, some scientists involved with the project were glib. A month after the accident they dubbed a draft plan to return the evacuated islanders to their home Project Hardy, as in Thomas Hardy, who wrote *The Return of the Native*.

will be born in future generations." Thousands signed, but millions began to fear the specter of worldwide radiation. Public opinion was turning against hydrogen bomb testing.*

Teller and his allies insisted that there was nothing to fear from a little extra radiation, even as nuclear tests were strewing fallout around the globe. The AEC's Willard Libby declared to a university audience in 1956, "It is possible to say unequivocally that nuclear weapons tests as carried out at present do not constitute a health hazard to the human population." He was lying. One test in 1957 produced "observable fallout on Los Angeles." And worldwide, strontium-90 levels were indeed rising rapidly. Scientists gathered data from unusual places. A research group in St. Louis pushed for mothers to donate 50,000 baby teeth for analysis. Others sampled the bones of children who died of other causes.† All the data showed that concentrations of strontium-90 were doubling every two years.

Teller, for his part, also tried consistently to squelch the growing fears about fallout. The radiation from atomic testing is "very small," he argued. "Radiation from test fallout might be slightly harmful. It might be slightly beneficial." He ridiculed the public's concerns. Afraid of the risk of mutations caused by radiation? "Our custom of dressing men in trousers causes at least a hundred times as many mutations as present fallout levels," he wrote in 1962, "but alarmists who say that continued nuclear testing will affect unborn generations have not allowed their concern to urge men into kilts." Teller even suggested that the dead captain from the *Daigo Fukuryu Maru* might have died from

---

* Pauling received his second Nobel Prize—the peace prize—for his efforts to warn the world about the danger of fallout. In his presentation speech for Pauling's award, the chairman of the Nobel committee noted, "The opposition Pauling encountered came first of all from two scientists, E. Teller and W. F. Libby, of the U.S. Atomic Energy Commission."

† At a secret 1955 meeting Libby proposed a particularly gruesome way to get samples to measure strontium-90 concentrations: "So human samples are of prime importance," he said, "and if if [*sic*] anybody knows how to do a good job of body snatching, they will really be serving their country."

hepatitis, not from radiation exposure.* In his view, the "fallout fear-mongers" were damaging the security of the United States because they were threatening to end his nuclear schemes. In Teller's view, "insignificant and doubtful medical considerations" about fallout led to an event "which has contributed decisively to our weakness and our danger": a nuclear testing moratorium.

In March 1958, Nikita Khrushchev came to power in the Soviet Union. Within days, he took the offensive against the United States. "The Administration was bracing itself today for Moscow's next big propaganda strike," warned the *New York Times* on March 29. "It is expected to be a declaration that the Soviet Union would end nuclear testing or production or both." Two days later, the plan was revealed: a complete moratorium on nuclear testing. On Moscow Radio, Andrei Gromyko, the foreign minister, announced the "cessation of tests of all forms of atomic and hydrogen weapons in the Soviet Union." The world wanted a solution to the growing fallout problem and a stop to the nuclear arms race, and the Soviet Union, unlike the United States, had responded. With the promise to suspend testing, "Russia has beaten us on propaganda all around the world," declared House Speaker Sam Rayburn.

This immediately posed a problem for U.S. politicians. How should they respond to the USSR's moratorium on nuclear testing? Should they ignore it and risk losing ground in the propaganda war against their Communist rival, or should the United States also cease testing? Teller was dead set against such a ban. Stopping nuclear testing was tantamount to surrendering America's nuclear advantage to the Russians. Teller would do almost anything to stop it from happening. At an

---

* Teller's arguments defending nuclear testing ranged from the disingenuous to the downright outrageous. At one point he attacked scientists who had the temerity to suggest that fallout-caused mutations might be a bad thing: "Deploring the mutations that may be caused by fallout is somewhat like adopting the policies of the Daughters of the American Revolution, who approve of a past revolution but condemn future reforms."

Atomic Energy Commission meeting in May, Teller argued that the United States needed a combination of underground and surface testing to get necessary data on new weapons systems. He warned that banning even surface explosions, much less following Russia's lead and banning tests entirely, would prevent the development of antimissile warheads. Then he stressed that a moratorium would damage Project Plowshare, his program for peaceful nuclear bombs.

Despite Teller's arguments, the pressure was too great for the administration to resist. The United States would follow the Russian lead—it would voluntarily cease testing nuclear weapons right after performing a last (and hastily cobbled together) test series. On November 1, 1958, nuclear fires stopped blazing in the United States.

Teller believed the moratorium was a huge mistake. He felt that with the ban in place the United States was getting weaker and the Soviet Union was getting stronger, and he repeatedly accused the USSR of cheating on the moratorium, of testing weapons underground or in space during the temporary test ban.* He concluded that U.S. adherence to the test ban had squandered the American nuclear advantage, and that his country was woefully unprepared for a coming "limited war" with the Soviet Union, a limited war that would almost certainly include the use of nuclear weapons. And developing new nuclear weapons required nuclear testing.

The mask had come off. Teller's opposition to the test ban had little to do with a vision of a fusion utopia. His future was not a future of peace, but of war. He had tried to stop the test ban because he wanted the United States to be prepared for tactical nuclear war with the Soviet Union. He had used the promise of peaceful nuclear explosions as a tool to ensure continued military research—and to make sure that weaponeers had more bombs to design. Teller's Plowshare was not

---

* There was no evidence for this. Teller was consistently (and unreasonably) pessimistic about the ability to detect Soviet nuclear tests underground and in space, a striking contrast to his consistent optimism about fusion weapons and his other nuclear schemes.

a vision from the prophet Isaiah, but one from the prophet Joel: "Beat your plowshares into swords, and your pruninghooks into spears: let the weak say, I am strong."

———

As it turns out, the test ban was only a temporary inconvenience to Teller. Throughout the moratorium, he made plans, so he could quickly resume his work when the agreement finally fell apart. Through 1959 and 1960, even as the moratorium held, he pushed Project Chariot—the Alaskan harbor—despite increasing resistance from the locals. When the moratorium ended after a Russian test in August 1961,* Teller was ready.

Within weeks, the United States was testing again, above and below the ground. The first new test series, Nougat, began on September 15. The seventh shot of Nougat, code-named Gnome, was a Plowshare test of nuclear excavation. A relatively small nuclear bomb, 3.1 kilotons, was placed deep in a shaft in a salt dome. When it went off, it instantly vaporized more than two thousand tons of rock and created a huge spherical cavern about 160 feet across. (It also vented a plume of radioactive smoke and steam, even though the radioactivity was supposed to be entirely contained.) The next Plowshare test came a few months later, in July 1962. It was the first shot of Operation Storax. Code-named Sedan, it used a 100-kiloton warhead buried 600 feet underground. Sedan carved an enormous crater—1,200 feet across and 300 feet deep—into the Nevada landscape. (Sedan, like Gnome, spewed radioactive ash into the air.) It was exactly the sort of test Teller needed: he proved that fusion weapons could move earth on a huge scale.

Unfortunately for Teller, shortly after Sedan the Alaska harbor project slipped out of his grasp. Local opposition was too great for the

———

* The United States accused the Soviet Union of reneging on its word. The USSR, on the other hand, had declared that the moratorium would only be valid so long as Western countries all ceased testing, and France exploded its first nuclear bomb in 1960. From the Russians' point of view, this ended their commitment.

Atomic Energy Commission to overcome, and in 1962 the plan was scrapped. But that did not stop Plowshare from marching forward. If anything, it gained momentum. In May 1962, even before the Sedan test was complete, President Kennedy ordered the AEC to tackle the problem of building a second Panama canal with fusion devices.

Such plans were becoming increasingly harder to make, though. During the moratorium and after it, Presidents Eisenhower and Kennedy had been negotiating a formal test ban treaty with Khrushchev. The so-called Limited Test Ban Treaty was signed in 1963; this was the agreement that banned anything but underground nuclear explosions. It also forbade any tests that allowed radioactivity to leak beyond national borders.

Teller bitterly fought the treaty. He and his allies attempted to undermine it at every turn. He did it directly, warning Congress of "grave consequences for the security of the United States and the free world" should they decide to ratify it: "you will have given away the future safety of our country." Teller also tried to sink the treaty indirectly, with Plowshare. Throughout the negotiations, the AEC kept trying to build a loophole into the treaty's language saying that peaceful nuclear explosions—Plowshare—should be exempt. Every American draft of the treaty had that exemption written into it. The Russians were against the exemption; they countered that "peaceful" nuclear bombs were "superfluous and even dangerous." After years of negotiation, Kennedy gave up on the exemption, and the accord was signed. However, it was only a matter of months before both sides were violating the brand-new agreement.

When, in January 1965, American planes first sniffed the cloud of radiation drifting over Japan, they had little idea that it was the first test of Project No. 7, the Soviet answer to Project Plowshare.* A Sedan-like explosion, 140 kilotons' worth, carved the crater that became Lake Cha-

---

* Interestingly, Teller had warned of a coming "Plowshare gap" in 1962. "The Communists might develop Plowshare before we do," he wrote. "The time may be near when the Russians will announce that they stand ready to help their friends with gigantic nuclear projects."

gan. The radiation that was released by the blast violated the new treaty, as did the radiation that vented from the April 1965 Palanquin shot in Nevada. Palanquin was a Plowshare test meant to see how nuclear weapons cratered dry rock, material similar to what would be encountered in digging a second Panama canal. (The radioactive plume from Palanquin rapidly crossed the northern border of the United States into Canada and beyond.) Soon the Americans and Russians were accusing each other of violating the new treaty. Peaceful nuclear explosions were not exactly maintaining the peace.

Despite all the effort and money poured into peaceful fusion explosions, the United States never got any nonmilitary benefit from Project Plowshare. No gigantic earth-moving projects ever materialized; neither the Alaska harbor nor the second Panama canal ever got beyond the planning phase. (President Nixon formally abandoned the latter project in 1970.) Of all the marvelous applications suggested by Teller and his colleagues, only one was actually tested. Three nuclear tests carried out in the late 1960s and early 1970s, code-named Gasbuggy, Rulison, and Rio Blanco, attempted to use nuclear explosions to release natural gas. (The theory was that a big enough explosion would fracture the rocks trapping the gas.) But the three tests were not terribly successful. Rio Blanco failed because the bomb didn't produce caverns of the expected shape. At first, Gasbuggy and Rulison seemed to work. The nuclear bombs shattered rocks around the test site and natural gas poured out of the wells. Unfortunately, the gas was radioactive, and no utility would buy it. After twelve years of trying and twenty-seven nuclear tests, Project Plowshare sputtered to a halt without ever having proved the usefulness of peaceful nuclear bombs.

Thirty years after Teller first dreamed of liberating the power of the sun upon the Earth, Project Plowshare was dead. Even the discovery of oil in Alaska in the late 1960s didn't make his proposal of a bomb-carved harbor any more palatable. In his waning years, Teller turned away from peaceful nuclear explosions and back toward using fusion as a tool of war, dreaming up unworkable schemes to defend the United States from a Soviet missile attack. He was behind President Ronald Reagan's infamous "Star Wars" program, which, in its first incarnation, would have seeded the heavens with fusion bombs. If the Communists launched their missiles, Teller's orbiting bombs would detonate, shooting out beams of x-rays that would destroy the incoming warheads. The project was abandoned as unworkable after just a few years, another example of Teller's manic optimism.

Throughout his career, Teller schemed and plotted to prevent any sort of hiatus in his quest for nuclear supremacy. Treaties with the Soviet Union were signs of weakness; detente and peacemaking would just lead to the destruction of America. Project Plowshare was a lie; to Teller, it was not a tool of peace but a means to undermine treaties with the Soviet Union. Teller was a man of swords, not plowshares. "I've never seen [Teller] take a position where there was the slightest chance in the interest of peace," said Isidor Rabi. "I think he is the enemy of humanity."

———

Project No. 7 had a little more success than Project Plowshare. After the creation of Lake Chagan, the Soviets briefly experimented with nuclear excavations of lakes and dams, but the results were disappointing. The Russian efforts to turn on gas and oil wells with bombs were more successful than the American tests. Production often increased dramatically. But reports indicate that at least one oil field is contaminated with radioactivity, and its oil is "not acceptable to regional refineries."

In 1966, the Soviets used a nuclear bomb to shut off a gas well and snuff a runaway fire. They also used nuclear explosives to make underground caverns for storing toxic waste, to break up mineral ores, and to create seismic shockwaves to aid in the exploration for natural resources. Sometimes accidents happened: the Kraton-3 explosion, a seismic experiment, vented so much radioactive steam into the Siberian tundra in 1978 that the Soviets had to declare a two-kilometer exclusion zone around the site.

All in all, Project No. 7 consisted of 122 nuclear explosions between 1965 and 1988. Their results were mixed at best. Hydrogen bombs, it turned out, did not give humanity the power to move mountains or to reshape the landscape to suit its fancy. What they provided was a far cry from a fusion-crafted utopia.

Fusion bombs were just that—bombs. They were swords too crude to be shaped into plowshares, unable to benefit humanity in any tangible way. Scientists would have to come up with entirely new ideas if they wanted to harness the power of the sun without getting burned.

CHAPTER 4

KINKS, INSTABILITIES, AND BALONEY BOMBS

Among other bodies which the alchemists of the middle ages thought it possible to discover, and accordingly sought after, was a Universal Solvent, or *Alkahest* as they named it. This imaginary fluid was to possess the power of dissolving any substance, whatever its nature, and to reduce all kinds of matter to the liquid form. It does not seem to have occurred to these ingenious dreamers to consider, that what dissolved everything, could be preserved in nothing.

—GEORGE WILSON, *RELIGIO CHEMICI*

The sun itself needs no bottle. It is held together by its own gravity; the mutual attraction of all its atoms is able to keep the fusion engine in its belly from blowing itself apart. But any lump of material smaller than a star does not have enough gravitational force to counteract the enormous pressure of an expanding fusion reaction. For humans to succeed in making an earthbound sun, scientists would have to figure out how to contain the fusion reaction with an external force—figure out how to bottle it up.

The Teller-Ulam design used atom bombs to create a temporary bottle. Pressure from the radiation of a fission bomb squashed the fuel from the outside; pressure from the explosion of a fission "spark plug" compressed the fuel from the inside. Caught in between these two nuclear anvils, the deuterium and tritium fuel was crushed, heated, and bottled up for a fraction of a second. The result was a brief burst of fusion energy: an exploding sun. However, the brevity and violence of the explosion made it suitable only as a weapon of war. To harness the power of fusion for peaceful purposes, scientists needed a much subtler kind of bottle.

In the early 1950s, the need was growing urgent. In the past, the United States had always produced more energy than it needed, but that trend was rapidly changing. Economists and scientists knew that by the end of the 1950s, America would have to begin importing fuel—oil—to keep its economy going. In Britain, the situation was even worse; dependent on oil imports, the United Kingdom was embroiled in a spat with its main supplier, Iran.* The West was getting its first taste of oil addiction, and it wasn't pleasant. Fusion energy—if scientists could design a bottle to contain it—could prevent a future where the Western world was kept hostage to a dwindling and increasingly expensive supply of foreign oil.

On March 25, 1951, Argentina announced that it had designed such a bottle. Argentina's scientists were claiming they had solved humanity's energy problems. It was a few days before the Greenhouse tests, and Ivy Mike was months away. The United States had not yet liberated fusion energy, but Argentina's president, Juan Perón, was gleefully bragging about having generated "thermonuclear reactions" and harnessing the power of the sun.

The *New York Times* reported the claim on page 1: "The project is still in the early stages, but when Argentina is able to produce as

---

* In early 1951, the new Iranian prime minister, Mohammed Mossadegh, tried to nationalize Iranian oil. This was unacceptable to Britain and to the United States. Declared *Time*'s man of the year for 1951, Mossadegh was ousted in 1953 by a coup. Not surprisingly, the coup was engineered by Britain and America in a CIA operation known as TPAJAX. Out of the ashes was born a new company: British Petroleum.

much energy as deemed necessary, all will be used solely for industry, President Perón declared." Not only was Perón willing to forswear his bomb-making ambitions, but he couldn't resist tweaking the scientists in the United States and the USSR who were trying to turn fusion energy into weapons. "Foreign scientists will be interested to learn that while working on the thermonuclear reactor, the problems associated with the so called hydrogen bomb were studied in great detail, and we have been shocked to find that results published by the most reputed experts are far removed from reality," Perón told a gaggle of Spanish-language reporters who had been summoned to a press conference at the Argentine presidential palace.

For such a dramatic claim, Perón provided very few details. The reactor used "a totally new way of obtaining atomic energy"—fusion. Experiments had been under way for some time and had yielded some very promising results, bringing matter to temperatures of "several million degrees." This success led Perón to establish a pilot fusion energy plant to create "artificial suns on earth." This plant was on a small island in a lake near the Chilean border: Huemul Island.

The director of the Huemul reactor was a German-speaking scientist named Ronald Richter, of whom little was known. After Perón finished speaking, Richter addressed the Spanish-speaking reporters through a translator. "What the Americans get when they explode a Hydrogen bomb, we in Argentina achieve in the laboratory and under control," Richter said. "As of today, we know of a totally new way of obtaining atomic energy which does not use materials hitherto thought indispensable." At the press conference and in a follow-up one the following day—no foreign press allowed—Richter spoke of controlled explosions of lithium and hydrogen and deuterium. And he claimed that he had achieved fusion at his mysterious lab on Huemul Island: "Yes, sir, for the very first time a thermonuclear reaction has been produced in a reactor."

The statement promptly set off a firestorm. Many in the physics community quickly rejected the claim; scientists around the world scoffed at Richter. When American reporters asked David Lilienthal whether there was the "slightest chance" that Argentina had attained fusion, he answered, "Less than that." A Brazilian scientist noted that "It is strange that the names of eminent physicists working at present in Argentina are not associated with the announced atomic project." And when asked what material other than hydrogen Richter could be fusing, a former Manhattan Project physicist, Ralph Lapp, was quick with an answer. "I know what that other material is that the Argentines are using," he told *Time* magazine. "It's baloney." (*Time* promptly dubbed Richter's reactor the "Baloney bomb.")

Perón bristled at the criticism. "I am not interested in what the United States or any other country in the world thinks," he snarled, lashing out at the foreign politicians and newspapers who "lie consciously" and spread deception. "They have not yet told the first truth, while I have not yet told the first lie."* And even as some scientists ridiculed Perón's claims, others began to chime in, supporting Richter's assertions. The mystery of Huemul Island, a drama that would last for months, was getting deeper by the day. It was just beginning.

Richter's claims marked the official beginning of a quest that had been in the planning stages for a long time—the quest to liberate the energy of fusion for the benefit of mankind. Years before scientists achieved fusion on Earth, they had realized that the uncontrolled violence of hydrogen bombs was far from an ideal way to harness the sun's power. What physicists really wanted was a fusion reaction they could control. They wanted a reactor that produced energy by fusing hydrogen into helium, and they wanted it to be stable, unlike the dangerous evanescent explosion of a fusion weapon. To create a workable reactor that would tap the unlimited potential of fusion energy, scientists needed to build a sun in a bottle.

Soon, scientists the world over were squabbling, alternately claiming triumphs and debunking them. The Huemul drama was the first act in the quest to create a tiny, controllable fusion reaction. But it was far from the last.

---

* The *New York Times* reporter wryly noted that "There seemed to be some puzzlement, however, over the President's declaration that, while foreigners systematically lie, he always is a shining champion of the truth."

At first glance, it seems impossible to make a bottle sturdy enough to contain a burning sun. What kind of material is strong enough to hold a fusion reaction? To get even the most fusion-friendly atoms to stick to one another, they have to slam together hard enough to overcome their mutual electric repulsion, so the atoms have to be extraordinarily hot—tens or hundreds of million degrees Celsius.* But matter at such high temperatures is very hard to contain. It is hotter than anything on Earth, far hotter than the melting point of steel. Even a diamond vessel would instantly evaporate in temperatures that extreme. Million-degree substances act almost like universal solvents, eating through whatever substance you put them in. Nothing on Earth would be able to contain such hot matter, at least not without some extraordinarily clever tricks.

Ronald Richter did not have the credentials one would expect of someone who could come up with such a clever trick. He didn't have a terribly strong scientific background. As a student, he apparently had tried to study (nonexistent) "delta rays" coming from the earth, but his proposal was rejected by his professors. His adviser merely remembered him as a "so-so" student. He hadn't published any scientific papers and had scant experience in the laboratory. But when Perón suddenly announced that Richter had created a sun in a bottle, he caught the world's attention.

In the days after Perón's announcement in March 1951, Richter provided a few more details about his reactor, which he called the "thermotron." He described the device as a "solar reactor furnace" and said the reactor worked by fusing deuterium with lithium—a light metal whose atoms have three protons and three or so neutrons—at 10,000 degrees Fahrenheit. This was far short of the tens of millions of

---

* Or Kelvin. Or even Fahrenheit, if you prefer. At these sorts of temperatures, it scarcely matters.

degrees that fusion scientists thought were required to initiate such a reaction. Richter also said that the reactions in the thermotron created little explosions, micro-fusion-bombs, which, however, were well-contained by large stone walls that surrounded the furnace. For most scientists, this announcement only increased their skepticism. But for others, Richter's work began to seem plausible, and they started jockeying to share in the credit for the discovery.

On April 1, the *New York Times* announced that a French physicist supported Richter's claims. The physicist was asserting that, a few months prior, he had performed experiments whose results bore a "striking similarity" to what the Argentine scientist was seeing. In the same issue, the *Times*'s science editor, Waldemar Kaempffert, wrote that "Richter admits that his process is not new," and the journalist listed some of his intellectual forebears: the Britons John Cockcroft and Robert Atkinson; the German Fritz Houtermans; the Russian émigré George Gamow; and, of course, Edward Teller. Kaempffert conceded that Richter might have made a breakthrough, but he rejected Perón's comment that everyone else was on the wrong track. "American and European scientists are fully aware of the work of Atkinson, Houtermans, Gamow, and Teller," he sniffed. If Richter had made a breakthrough, it was not Argentina's alone. It was due, in part, to the work of American, British, and German physicists.

Later that month, Perón pinned the Peronist loyalty medal on Richter's chest, and in May he established a national physics laboratory to exploit the discovery. The Dutch government started negotiating nuclear research deals with Argentina. The South American country was trying to become a major player in nuclear politics. At the same time, though, the criticism and scorn from foreign skeptics became increasingly caustic. *Time* magazine and other outlets picked up a rumor that Richter had been arrested. A Brazilian newspaper, *Time* reported,

said that Dr. Ronald Richter, the former Austrian scientist, was arrested after technical experts of the Argentine army had discovered that Richter "was not sufficiently advanced as a physicist" to achieve the atomic release Perón had claimed. Three experts informed Perón that Richter, in their opinion, was nothing more than a "colossal bluff."

It wasn't true, but it amplified the claims of fraud that surrounded Richter. Also in May, the Austrian physicist Hans Thirring published an article that asked whether Richter's scheme was "a swindle." The answer consisted of the following possibilities:

(a) Perón has fallen victim to a crank suffering from self-delusion   50%
(b) Perón has been taken in by a sly swindler   40%
(c) With the aid of Richter, Perón is attempting to bluff the world   9%
(d) Richter's assertions are true   1%

Richter lashed back the next month—in the paranoid and combative style of the professional crank. "The reactor operation crew and I are deeply sorry for Herr Thirring, because he revealed himself to be a typical text book professor with a strong scientific inferiority complex, probably supported by political hatred," he wrote. And of the reports that he had been arrested in secret? "It must really have been the deepest degree of secrecy because I only know of it through the newspapers," Richter sneered. "I am not impressed by these well-known methods of psychological warfare." His research would continue unabated, and he would likely have a thermonuclear reactor "in full-scale operation in about ten months or so."

The rumors of Richter's arrest, at least, were false, and work continued at Huemul, apparently on schedule. In October, Richter announced that a large-scale experiment had been successful. By December, he was bragging that he "would be able to make convincing new demonstrations within three months." The Associated Press reported Richter's claim that he was in negotiations with a "highly industrialized foreign country" to trade his nuclear secrets for money and raw materials. He added that foreign skeptics "soon would have to eat their words."

Instead, Richter was the one who would get his comeuppance. Three months later, he failed to produce his "convincing new demonstrations," and he became the butt of jokes in Argentina. Detractors, seeing Richter in a café with a bandage on his hand, commented snidely that the good doctor had been wounded when one of his atomic bombs exploded in his hands. In the lab, Richter's behavior was becoming increasingly bizarre. He requested pumps to inject gunpowder into the reactor. In April 1952, Pedro Iraolagoitia, a Peronist navy pilot, visited Huemul to inspect the plant. He was shocked when Richter deliberately blew up a tank of nitrogen and hydrogen, blowing the door to the lab clean off. Weirder still, right after he triggered the explosion, Richter scuttled over to his instruments and on a piece of paper spewed out by one piece of equipment, he wrote "atomic energy." Iraolagoitia figured that Richter was insane. Soon he had convinced Perón to launch an investigation into the Huemul project, and a group of scientists and politicians visited the island in September 1952.

The visit was a fiasco. Richter showed the scientists his fusion reactions. In the reactor chamber, vivid red lithium and hydrogen flames spewed forth; the dials of the Geiger counters fluttered. The scientists weren't impressed. Neither were the gamma-ray detectors that the physicists brought with them. Unlike Richter's Geiger counters, the gamma-ray detectors showed no evidence of radiation whatsoever—radiation that had to be there if, indeed, fusion was occurring. Stranger still, when they exposed Richter's equipment to a real source of radiation—a piece of radium—the counters didn't chirp at all. The equipment had been rigged. The Geiger counters weren't responding to radiation but to the electrical discharge that ignited the ersatz fusion reaction. At best, Richter was deluded. At worst, he was a fraud. Even Perón, Richter's biggest backer, had to admit that Richter's nuclear fusion was a farce.

The failed Huemul project quickly became an embarrassment for Perón. Richter began to face accusations from Argentina's legislators, and in December, despite official denials, the Argentine dream of fusion energy was over. Rumors of Richter's incarceration began to appear in the press, but in truth it was two more years before he was finally arrested.

Perhaps it was naïveté or optimism that drove him; perhaps it was greed. Perhaps it was the desire for power. Whatever the reasons for Richter's bold claims, he had wasted millions of dollars' worth of Perón's money in his pursuit.* If he had succeeded, Perón's Argentina would have solved the world's growing energy crisis overnight. Humanity would have considered Richter its savior.

Instead, Richter found himself accused of fraud, scorned and humiliated. He was just the first casualty of the quest to put the sun in a bottle.

———

Lyman Spitzer, a physics professor at Princeton University, was about to leave for a ski trip to Aspen in March 1951 when the papers broke the story about Richter's fusion reactor. Spitzer was instantly incredulous, but at the same time he was intrigued. On the ski slopes, he wondered just how to make a device that could hold a miniature sun. By the end of his sojourn, Spitzer was well on his way to designing one. Instead of using nuclear weapons to contain fusing hydrogen, Spitzer would exploit the odd properties of extremely hot matter, properties of what physicists call a *plasma*.

Heating an object changes the way it behaves. A frozen hunk of water is solid ice. Put it on a table and it will retain its shape. Heat it a bit and the ice changes, melting into a liquid. Though liquid water still has a definite volume, it no longer has a fixed form; it will change its shape to fit whatever container you put it in. Heat it some more and the fluid changes again. The water boils into a gas: steam. As a gas, the formless cloud no longer even has a definite volume. A gas expands or contracts, depending on the pressure and temperature of its surroundings. As matter gets hotter and hotter—more and more energetic—its

---

* Just how much is a subject of debate. Contemporary reports put the cost of Huemul as low as $3.7 million and as high as $70 million. Of course, the international humiliation was priceless.

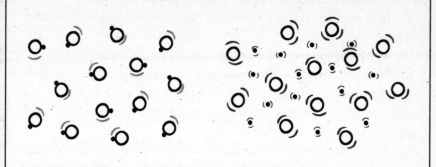

GAS VERSUS PLASMA: In a gas (left), every electron is stuck to an atom. In a plasma (right), the electrons roam free, attracted by nuclei, but not attached to any single nucleus.

atoms' random dancing speeds up and it eventually changes from solid to liquid to gas. This much scientists knew for centuries. Only at the end of the 1800s did physicists begin to realize that extremely hot gases changed their properties yet again. The reason has to do with the composite nature of the atom.

As scientists realized at the beginning of the twentieth century, atoms are not quite as uncuttable as their name would imply. Protons and neutrons sit in the center of the atom, making up a small, dense, heavy, positively charged nucleus. Surrounding the nucleus are light, negatively charged electrons. Ordinarily, electrons are bound to a nucleus; the opposite charges of the protons and electrons attract each other, so the electrons cannot easily shake free. In fact, the nucleus and the electrons attract each other so strongly that the whole mess behaves very much like a single object.

Yet this isn't always the case. Raise the temperature and the atoms start moving faster and faster. There's a lot of energy about, and some of that energy winds up exciting the electrons. If the temperature is hot enough, the electrons get so excited, so pumped full of energy, that they can escape the bonds of their nuclei, and the atom loses an electron. As the temperature rises, the available energy increases, the electrons become more excited, and one by one they escape their bonds.* Finally, at a high enough temperature, all the electrons are stripped from their nuclei.

The electrons are still nearby, unattached to any particular nucleus. Unbound electrons and nuclei roam in one big blob, unattached to each other. At extremely high temperatures a hunk of hot matter becomes an undifferentiated soup of unconnected negatively charged electrons and positively charged nuclei.

This is a plasma. Pour enough energy into a piece of matter—heat it enough—and atoms lose their individuality. The positively charged nuclei are still attracted to the negatively charged electrons, but they are not bound together. And this gives a plasma some unusual properties. Unlike most kinds of ordinary matter—unlike most solids, liquids, and gases—the free-floating electrons and protons of a plasma are strongly affected by electric and magnetic fields.

To Lyman Spitzer, this suggested a design of a bottle that could hold a miniature sun. Spitzer's bottle would not be made of steel or stone or diamonds. It would not be made of any kind of material at all; after all, nothing would be able to stand up to the immense heat of a fusion reaction. Spitzer's bottle would be made of invisible lines of force: it would be made of magnetic fields.

By the twentieth century, these fields were extremely well understood. Physicists had long been amazed by the intricate interplay of electric fields, charged particles, and magnetic fields, but in the nineteenth century, physicists figured out that these interactions are governed by only a handful of relatively simple rules. Nonetheless, even simple rules can have seemingly complicated consequences.

For example, the laws of electromagnetism dictate that moving charges are affected by magnetic fields, while stationary ones are not. It's a quirky-sounding rule, but it's what the equations dictate: if you put a

---

* It is somewhat analogous to how the force of the stem that binds the apples to the branch of an apple tree gets overwhelmed if you shake the branch energetically enough. Pour enough energy into the branch and you will break the bonds, freeing the apples.

stationary charged particle (like a proton) in a magnetic field, it won't feel the field at all. A charged particle that is moving, on the other hand, is tugged and deflected by a magnetic field. More specifically, a moving charged particle feels a magnetic pull perpendicular to its motion. This force makes the particle change course. Instead of moving in a straight line, the particle moves in a circle, and the stronger the magnetic field, the tighter the circle. Conversely, the equations of electromagnetism dictate that a moving electric charge (like an electron moving down a wire) will generate a magnetic field. A stationary electric charge won't. In mathematical terms, these are pretty simple rules to describe. But just these rules can give you a hint of how complicated a plasma must be.

In a plasma, you have a large number of charged particles—electrons and nuclei—moving about at relatively high speeds. These moving particles generate magnetic fields. These magnetic fields change the motion of the moving particles. When the motion of the moving particles change, so do the magnetic fields that they are generating—which changes the motion of the particles, changing the magnetic fields, and so on. Add to that the electric attraction that the electrons and nuclei feel for each other and you've got an incredibly complex soup.

Nevertheless, to Spitzer, the mere fact that the plasma responds to magnetic fields suggested a way to bottle it up. He realized that if you had a plasma moving through a tube and you subjected that tube to a nice, strong magnetic field in the proper orientation, the charged particles in the soup would be forced to move in little circles. They would spiral down the tube in tight little helices, confined by the magnetic field, never even getting close to the walls of the cylinder. The plasma would be confined. In theory, even an extremely hot plasma could be trapped in such a bottle. Furthermore, it was fairly easy to generate the right sort of magnetic field: just wrap a coil of wire around the tube and put a strong current through it; the moving charges in the wire create just the sort of field that is needed. It was a simple, but powerful, idea.

The only problem with Spitzer's tube was that it bottles the plasma on the sides, but not at the front or the back of the tube. When the moving plasma reaches the end of the tube, it spills right out. So what to do? Spitzer's next clever idea was to imagine a tube without end: a donut. Such a donut (or a *torus*, as physicists and mathematicians like to call it) is just a tube that circles back upon itself. The plasma would move around and around in a circle, never spilling out the end of the tube. With the right magnetic field, it would be a perfect bottle.

Unfortunately, the very curvature that gets rid of the end of the tube makes it very difficult to set up the right kind of magnetic field. The straight tube merely needed a wire curled around it. But when you bend that tube into a donut, the loops of wire on the inside of the donut get bunched up and those on the outside get stretched and spaced out. As current flows down the wire, the magnetic field is stronger where the loops of wire are close together—near the donut hole—and weaker at the edge where the loops are far apart. The nice, even magnetic field in the tube is destroyed; it becomes an uneven mess with a strong side and a weak side. This unevenness is a big problem: it causes the nuclei and electrons in the plasma to drift in opposite directions and into the walls of the container. The bottle quickly loses its contents. It leaks. A straight tube leaks out its ends; a curved tube leaks through its sides.

Even though a torus-shaped bottle would be leaky, Spitzer quickly came up with a design to minimize the leak. Instead of a simple donut, he reasoned, it would be better to have two half donuts. These half donuts would be connected by tubes that crossed each other: a figure eight. The half-donut sections have the same problem as the full-donut bottle: the electrons and nuclei drift in opposite directions. However, because the tubes cross each other, the plasma winds up going through one half donut clockwise and the other one counterclockwise. This means that the drift on one side should be cancelled by an equal and opposite drift when the plasma goes through the other half donut. It doesn't *quite* work that way; the drifts don't cancel exactly, and the plasma still leaks out a bit, but the leak isn't quite as severe as it would be in a torus-shaped bottle.

The figure-eight bottle was a good enough design for Spitzer to begin experimenting. In July 1951—as the Huemul controversy continued raging in the press—Spitzer got a grant from the Atomic Energy Commission and set to work at Princeton, generating a serious design for a figure-eight-shaped fusion reactor: the Stellarator.

The Stellarator wasn't the only fusion reactor in town. Shortly after the Livermore laboratory was founded, some of its scientists proposed a slightly different shape for a magnetic bottle. They would stick with a straight tube. Instead of wrestling with the problems caused by curving the plasma's path, they would try to cap the ends of the tube by tweaking the magnetic fields slightly. Strong magnetic fields at the ends of the tube and slightly weaker magnetic fields at the center would create barriers that would behave almost like a mirror. Some—not all—of the plasma streaming to the end of the tube would be reflected back inside. This *magnetic mirror* was extremely porous, so it was clearly not a perfect bottle, but neither was the Stellarator. The Livermore scientists got down to work.

A third contender came from across the Atlantic. In the late 1940s, British scientists were also beginning to think about confining plasma, and their method relied on an entirely different phenomenon that they called the "pinch" effect.

A pinch starts with a cylinder of plasma. Since the electrons are free to move around inside the cloud, the plasma itself conducts electricity; it's almost like a piece of copper. You can send an electric current along a plasma cylinder just as you would along a copper wire. And just as in a copper wire, the current running down the plasma creates a magnetic field. But this magnetic field affects the particles in the plasma; it forces them toward the center of the cylinder. The current compresses the cylinder, crushing it toward its central axis. The stronger the current, the greater the effect, and the faster and tighter the plasma gets squashed. As an added benefit, the squashing heats the plasma. This is the pinch effect.

British scientists immediately seized on this effect as a way to confine and heat a plasma until it begins to fuse. Several British laboratories began work on pinch projects, particularly Oxford University's Clarendon, and neighboring Harwell, a few miles away. James Tuck, a physicist who had been involved with the Manhattan Project, worked briefly on the Clarendon fusion project before returning to Los Alamos, bringing the pinch idea with him.

To Tuck, the pinch method seemed an especially promising way to build a fusion reactor. If you could pump enough current into a plasma of deuterium and tritium, the plasma would heat and compress itself all in one fell swoop—perhaps enough to ignite fusion on a small scale. In 1951, Tuck asked for money to build a pinch machine, and in 1952 he built his first. In contrast to Spitzer's hubristic name—Stellarator implied a mini-sun—Tuck called his instrument the Perhapsatron.

The Perhapsatron, the Stellarator, and the magnetic mirror all showed great promise. At least on paper, they were all able to contain a plasma in a magnetic bottle. Within a few months, scientists had come up with not one but three containers for an uncontainable substance. The Atomic Energy Commission decided to pursue them all. By the time Richter was finally unmasked as a fraud, the United States had consolidated these three efforts into one project: Project Sherwood.*

At first, Sherwood's funding was modest, a few hundred thousand dollars or so per year. The budgets would not stay small for long. Though Project Sherwood was classified, it would soon hit the world stage. By mid-1955, rumors abounded that Britain, the USSR, and the United States were all trying to solve the world's energy problems with fusion—and that U.S. scientists were about to build a prototype fusion reactor. In August, fusion scientists from around the globe met in Geneva for the first UN Conference on the Peaceful Uses of Atomic Energy. The conference president, the Indian physicist Homi J. Bhabha, stunned the world with a bold pronouncement. "I venture to predict that a method will be found for liberating fusion energy in a controlled manner within the next two decades," he said. "When that happens, the

---

* Possibly so named because they were raiding the Treasury on behalf of "Friar" Tuck.

energy problems of the world will truly have been solved forever, for the fuel will be as plentiful as the heavy hydrogen in the oceans." The dream of fusion energy had been officially made public. Within twenty years, humanity would have limitless energy. The energy problems that had plagued civilization would be a thing of the past.

Lewis Strauss, the head of the AEC, was quick to claim a share of the dream. He confirmed that the United States was hard at work trying to build a reactor that would produce energy. The public and the press began to learn about Project Sherwood, if only the gross details. They knew nothing about the problems that were looming.

———

Fusion scientists started off very optimistic about their designs; on paper, the machines they were building seemed sure to work. In his 1951 proposal, Spitzer estimated his small Stellarator would generate about 150 million watts of power.* The Perhapsatron looked even more promising. It was technically simpler and thus seemed likely to achieve fusion sooner. Once it did, it would be easy to turn a pinch-type device into a reactor. It could behave like a fusion-powered version of an internal-combustion engine: inject fuel, compress it with a current, ignite it, extract the energy, and get rid of the nuclear "ash." It seemed almost *too* easy, and everybody was pursuing the idea. Despite the secrecy surrounding the early fusion reactor programs, U.S. scientists were certain the British and the Russians were working on pinch-type reactors.

The enthusiasm surrounding the technology, though, hid a lot of difficulties—and some infighting. Spitzer and the Princeton Stellarator team thought their idea was the path to fusion energy, and tried to tear down the Perhapsatron idea championed by their rival, Los Alamos's Tuck. In fact, Spitzer spent some of his AEC grant trying to prove that a Perhapsatron would not work. It was money well spent. Two of his

---

* This is about one quarter the power produced by a typical commercial power plant. Roughly speaking, a home consumes 1,000 watts on average, so 150 million watts would power 150,000 homes.

team members, Princeton professors Martin Schwarzschild and Martin Kruskal, found a very disturbing flaw that threatened to disrupt the Perhapsatron research program altogether. A pinched plasma was unstable.

Perhaps the easiest way to understand stability and instability is to imagine a ball sitting at the bottom of a hill. This is a stable system. Give the ball a slight nudge and it will roll right back to where it started. The system resists change; it won't be ruined by small perturbations. A ball perched on the top of a steep hill, on the other hand, is in a precarious position. Give it even the slightest nudge and it will roll down the slope, abandoning its previous place. This system doesn't resist change—indeed, even a tiny disturbance will change it dramatically. This is an unstable system.

Kruskal and Schwarzschild had discovered that a pinched plasma was like a ball perched on a hill. The slightest disturbance would destroy it. Send a current through a cylinder of plasma and it indeed squashes itself into a dense little filament of hot matter. But the filament is unstable. If it is not perfectly straight, if it has even the tiniest kink, the magnetic fields generated by the pinching current immediately exaggerate and expand the kink. This makes the kink grow, getting more and more pronounced. Any little imperfection in the plasma filament rapidly becomes a huge imperfection. In a tiny fraction of a second, the plasma kinks, bends, and writhes out of control.

As soon as the Perhapsatron started up in 1953, the Princeton team's calculations were proved correct. The Los Alamos experimenters found that as soon as they got a pinch, forming a nice, tight filament in the center of the Perhapsatron's chamber, it went poof! The pinch would disappear, setting the whole chamber aglow. High-speed cameras revealed the filament buckling and writhing, quickly striking the walls of the chamber. The *kink instability* had claimed its first victim. The Perhapsatron, as built, was incapable of fusing anything at all. The Los Alamos scientists needed to figure out how to stabilize the filament if they were to progress. They tried using an external magnetic field to

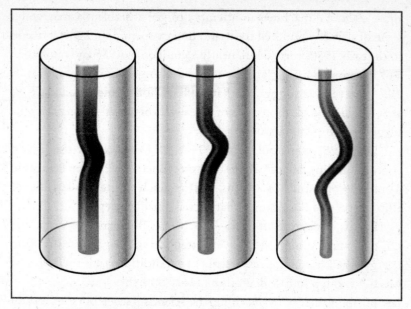

KINK INSTABILITY: If a pinching plasma has even a tiny kink in it, that kink will grow; the plasma will writhe out of control and hit the walls of its container.

"stiffen" the plasma filament somewhat, but the essential instability remained. Pinches were in trouble.

Soon, the other designs were as well. In 1954, Edward Teller figured out that a plasma held in place by magnetic fields was unstable under certain conditions. The magnetic fields behave somewhat like a collection of rubber bands: as the plasma pressure increases, they try to relieve the increasing tension by writhing. "They try to snap inward and let the plasma leak out between them," Teller wrote. This system was also unstable. Even a tiny irregularity in the magnetic field would rapidly get worse, and scientists would lose control of the plasma. The so-called Teller instability affected the Stellarator as well as Livermore's magnetic mirror approach. Instabilities were everywhere.

By the mid-1950s, all three groups had enormous difficulties to overcome. Their plasmas were unstable and their bottles were leaky. They spent ever-increasing amounts of money building bigger and more elaborate machines in attempts to get unstable plasmas under control. The few hundred thousand dollars spent on magnetic fusion in the early 1950s turned into nearly $5 million by 1955 and more than $10 million by 1957. Plans for reactors also got more ambitious: by 1954, Spitzer was suggesting that $200 million would buy a machine that would produce thousands of megawatts of power—bigger than the biggest power plants around.

Despite Spitzer's bold plans, Teller's Livermore got the largest share of funding, about half of the Project Sherwood money. Princeton came in second, and Los Alamos, with its pinch program, was a distant third. Yet it was Los Alamos that first claimed victory.

By the beginning of 1955—just before Bhabha's speech at the UN conference brought worldwide attention to the promise of fusion energy—the Los Alamos researchers saw indications that their plasma was hot enough to fuse deuterium. Every time they initiated a strong, fast pinch in their latest machine, the scientists saw a burst of tens of thousands of neutrons. This was very encouraging, because neutrons are the best indicator of a fusion reaction.

Everyone in the fusion community was hoping to achieve two main kinds of thermonuclear fusion in a reactor. The easier kind used a mixed fuel: deuterium and tritium. When a deuterium (a proton and a neutron) and a tritium (a proton and two neutrons) strike each other hard enough, they fuse, creating helium-4 (two protons and two neutrons). The remaining neutron flies off with a great deal of energy. So in a successful deuterium-tritium reaction, the products will be helium-4 and neutrons. Deuterium-deuterium reactions are a little more complicated; there are two ways this kind of fusion reaction tends to happen. As the two deuterium nuclei collide and stick, either a proton flies off (leaving behind a tritium nucleus) or a neutron flies off (leaving behind a helium-3 nucleus). These two *branches* of the reaction are roughly equally probable. Thus, if a reactor succeeds in fusing deuterium fuel, then the products will be helium-3, tritium, protons, and neutrons. Neutrons are produced by both deuterium-tritium and deuterium-deuterium fusion reactions. A burst of fusion—no matter whether the fuel is pure deuterium or deuterium mixed with

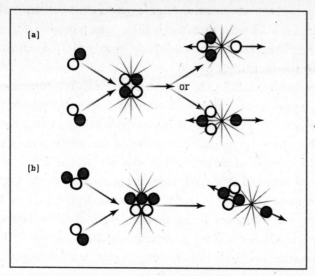

FUSION REACTIONS: (a) Two deuteriums collide and produce either a tritium and a proton or a helium-3 and a neutron. (b) A deuterium strikes a tritium and produces a helium-4 and a neutron.

tritium—will be accompanied by a corresponding burst of energetic neutrons.*

Despite the problem with the kink instabilities, Los Alamos scientists were optimistic. If they could make the pinch strong enough and fast enough, they thought, they could get fusion going before the kink instability destroyed the pinch. In fact, Tuck's calculations showed that such a machine could achieve breakeven—the fusion reaction in the machine would produce energy equal to what was needed to get the reaction going in the first place. (A fusion reactor that absorbs more energy than it produces is of no use to anyone.) And, Tuck argued, a larger machine could produce explosions equivalent to several tons of TNT per pinch. These explosions could be turned into usable energy,

---

* Making matters easier still, neutrons are easy to detect. Since they are neutral particles, they aren't affected by the magnetic fields of the plasma and zoom right out of the reactor vessel.

just as an internal combustion engine makes little fuel-air explosions turn a crank. Tuck built successively bigger pinch machines that could pinch the plasma harder and faster, and eagerly awaited neutrons produced by thermonuclear fusion.

When, early in 1955, the Los Alamos researchers turned on their newest, biggest pinch machine, Columbus I, they saw a burst of neutrons every time they pinched the plasma hard enough. Pinch. Neutrons. Pinch. Neutrons. No pinch, no neutrons. It seemed like a great success. From the number of neutrons they were seeing, the pinch scientists concluded they had attained fusion; the plasma inside the Columbus machine must have been heated to millions of degrees Celsius. But not everybody was convinced. Researchers at Livermore were skeptical that the pinch machine could reach the temperatures advertised. Thus, the plasma couldn't possibly be hot enough to ignite a fusion reaction. So where were the neutrons coming from?

The Los Alamos physicists started making careful measurements on their pinch machine to see if they could pin down the origin of those neutrons. To their chagrin, they soon discovered that the neutrons coming out the front of the Columbus machine were more energetic than the ones coming out the rear. In a true thermonuclear reaction, during which nuclei in a hot plasma are fusing with one another, the neutrons from the reaction should be streaming out in all directions with equal energy. This was not the case with Columbus, so, clearly, the Columbus neutrons weren't coming from thermonuclear fusion. They were coming from somewhere else.

The asymmetry provided a crucial clue. The scientists pinched the plasma by running a current through it. Neutrons that were flying out of the machine in the direction of the current had more energy than those that flew out against it. This revealed that the neutrons were the work of another instability. Just as a pinched filament is unstable when kinked slightly—because the kink grows and grows—it is unstable when a small section gets pinched a little bit more than the rest of the plasma. In this case, the small pinch grows progressively more pronounced; the plasma gets wasp-waisted and pinches itself off. The plasma begins to look like a pair of sausages. This is a *sausage instability*, and it creates some strong electrical fields near the pinch point. These fields accelerate a small handful of nuclei in the direction of the pinch current. These nuclei then strike the relatively chilly cloud of plasma and fuse, releasing neutrons.

From fusion scientists' point of view, this kind of fusion was worthless. Scientists were hoping to get a hot cloud of nuclei fusing with itself, a thermonuclear fusion reaction. Instead, Columbus had made a small handful of very hot nuclei interact with cooler ones. This was roughly equivalent to shooting nuclei at a stationary target, and doing that, scientists had concluded, would always consume more energy than it produced. The neutrons produced by the instability, dubbed *instability neutrons* or *false neutrons*, weren't a sign of energy production—just the opposite. Columbus's neutrons were the sign of energy consumption, not energy production. The false neutrons had given the Los Alamos scientists false hope. Even so, the pinch technique still seemed within striking distance of igniting fusion.

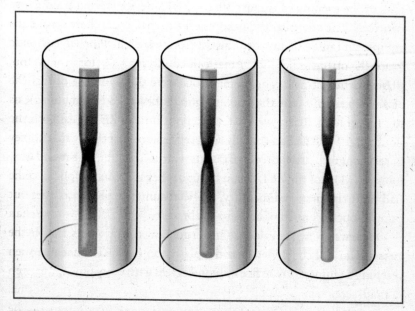

SAUSAGE INSTABILITY: If a pinching plasma is slightly narrower in one place, that narrowing will get more and more severe and eventually squeeze the plasma to make it look like a pair of sausages.

By this time, the Americans knew they had competition from both the Russians and the British. Project Sherwood poured increasing amounts of money into ever-larger machines of all types. The most expensive one in the Sherwood portfolio was the model-C Stellarator proposed by Spitzer, which would cost roughly $16 million to design and build. So it was a humiliation when it appeared that the British had won the fusion race with a much smaller and less-expensive machine: ZETA.

——

ZETA, which had cost less than $1 million to build, was a powerful pinch machine. Its name reflected the optimism of its designers; ZETA was an acronym for Zero-Energy Thermonuclear Assembly, thermonuclear because it would achieve fusion and zero-energy because it would produce as much energy as it consumed. It was a very bold claim.

ZETA began operation in mid-August 1957 at the Harwell laboratory near Oxford. It wasn't long before the machine made a big splash. Late in the evening of August 30, the ZETA device started producing neutrons. The scientists did hasty checks to make sure there wasn't an equipment failure of any sort; the neutrons were real. Pinch. Neutrons. Pinch. Neutrons. Like their American counterparts before them, the British physicists thought the neutrons were the signature of fusion; after all, neutrons were the smoking gun that everybody had been seeking for so long. There were a few skeptics on the ZETA team—some doubted that ZETA had actually achieved fusion—but the joyful chorus of self-congratulation drowned out the voices of doubt. The mood was jubilant. Most of the ZETA team thought they had finally done it; they had built the first, rudimentary, artificial sun. The physicists present popped open a bunch of beers to celebrate.*

After weighing the evidence and crunching the numbers, the physicists at Harwell concluded that the plasma in the ZETA machine was reaching a temperature of five million degrees with every pinch, creating

---

* The next morning, the puzzled cleaning staff, picking up all the bottles from the previous night's celebration, wondered aloud whether the ZETA machine had begun running on pale ale.

thermonuclear reactions and producing neutrons. If so, this was big news. It would be the first time that scientists had achieved fusion in a controlled environment. The British team naturally wanted to release their initial results right away, revealing the brilliant future of limitless energy to the public. But the Americans balked.

Earlier in the year, the British and Americans had decided to share data on fusion reactors with each other, and they were to decide jointly when and how to declassify the data and release it to the public. This last point became a source of contention. The Americans were reluctant to make an announcement about ZETA, in part because Project Sherwood had no signal achievements to brag about at the time. It looked as though the Brits had beaten the pants off the Yanks, so the Yanks needed some time to catch up. Thus, citing security issues, the United States tried to delay the announcement for a year. A second United Nations conference was scheduled for 1958, and what better place was there, U.S. officials argued, to release the results?

The British were unhappy about the American insistence on secrecy, and it is hard to keep a secret if one party wants to reveal it. Naturally, the secret didn't stay secret for very long. By early September, news of ZETA's success was leaking out. The English press was buzzing with rumors of successful nuclear fusion in the ZETA machine. By October, British scientists—including the Nobel laureate and Harwell lab head, John Cockcroft—hinted at encouraging results from the device. However, in deference to the Americans, nobody made an official pronouncement.

Hints about ZETA's success got harder and harder to ignore. In November, a spokesperson for the British Atomic Energy Authority (BAEA) stated, "The indications are that fusion has been achieved" at ZETA, but gave no further details. English scientists briefed the House of Commons on their achievement. But the weeks ticked away without any description of what, exactly, had happened at Harwell.

The mystery deepened in December. Even as the BAEA denied that the United States was deliberately "gagging" ZETA scientists, preventing them from releasing their results, a BAEA spokesman admitted that Britain was awaiting American approval to publish the details of the Harwell experiment. The British press was infuriated and accused the United States of playing politics with a crucial scientific achievement, of needlessly delaying publication of an important experimental result. The British had beaten the Americans at their own game, and Lewis Strauss and the Project Sherwood crew seemed to be holding up the declassification process to give themselves time to catch up to the Brits.*

To the ZETA scientists, it was more than merely frustrating. Without a publication, it was as if the experiment had never happened. In science, publication is everything; without it, an experiment is worthless. It is easy for anybody to make an outrageous scientific-sounding claim. If you use the right buzzwords, you can make it extremely convincing; you can easily make the public believe that your claim is true. That's what happened with Ronald Richter. Given a platform by Juan Perón, Richter trumpeted a remarkable achievement—based on pseudoscience—around the globe. Important people, including Perón, believed him. But very few scientists did. That is because Richter did not publish any scientific data that would have allowed specialists to verify his claim. To a scientist, an experiment is not believable without the precise details of how it was run and what the researchers involved observed. Only when scientists reveal the inner workings of an experiment to the world can their peers scrutinize the work and confirm or refute it. Only then will they be taken seriously. By blocking the publication of the ZETA results, the Americans were denying the British their chance at scientific glory.

Finally, the Americans succumbed to the pressure and gave Britain the go-ahead to publish the ZETA findings. When the Harwell scientists announced, in mid-January, that they were publishing their results in *Nature* at the end of the month, the British press was ecstatic. When they learned that the ZETA papers were to be accompanied by papers about Project Sherwood's pinch project, the press was absolutely livid. It looked as if Lewis Strauss and the Americans, with their expensive

---

* The American press was defensive. The *New York Times* rather lamely attributed the delay to the "backlog of experimental results" from fusion reactors that needed to be declassified.

machines, were trying to steal some of the Harwell laboratory's glory. "Admiral Strauss' tactics have soured what should be an exciting announcement of scientific progress so that it has become a sordid episode of prestige politics," blared the British Sunday *Observer*. Despite the hurt feelings, everyone was relieved that the long wait was about to end.

When the *Nature* papers finally came out on January 24, the British and American scientists held a joint press conference. John Cockcroft announced that it was "90% certain" that ZETA's neutrons had come from fusion, and outlined a twenty-year research plan that would lead to fusion reactors. The Americans presented their results, too, but they weren't nearly as striking as ZETA's. The press in the United States spun the story as a great British-American achievement. "Gains in Harnessing Power of H-Bomb Reported Jointly by U.S. and Britain," the *New York Times* declared; "Nations Called Equal—Many Questions to Be Resolved." America's Columbus II machine was given pride of place above Britain's ZETA, and the newspaper emphasized that the two nations were "neck and neck." However, the rest of the world's press ignored the American research and celebrated Britain's triumphant conquest of fusion energy. In England, tabloid papers blasted the news across their pages, promising "UNLIMITED POWER from SEA WATER": no more electricity bills, no more smog, no need for coal, power that would last for a billion years. Newspapers around the globe followed suit; they were quick to trumpet the prospect of limitless energy, energy that would be at humanity's fingertips within two decades. No longer would any nation be held hostage because of a lack of oil. Even the Soviets congratulated the British—pointedly ignoring the Americans—on their "achievement in harnessing thermonuclear energy" and expressed their "admiration."* ZETA seemed to have begun a new era of humanity, the era of unlimited fusion energy, and it was the envy of the world.

---

* Of course, the Russians tried to get a share of the credit, too. "British scientists pointed out that the ZETA installation used a method of thermo-isolation employing a magnetic field," read a dispatch in *Tass*, "which, as we all know, Soviet academicians I. E. Tamm and A. D. Sakharov first proposed in 1950."

Other nations began to emulate the British. The Swedes announced that they were building a ZETA-like device that could compete with the one at Harwell. Just two weeks after the announcement, Japanese scientists announced that they, too, had achieved thermonuclear fusion—and they were producing more neutrons than the British were. The Russians also started building a ZETA clone. But the Britons weren't going to fall behind: by early May, they were busy upgrading ZETA and were planning a more powerful (and more expensive, at $14 million) machine, ZETA II. Its designers thought that ZETA II would heat plasmas to one hundred million degrees and produce more energy than it consumed. It would be the world's first fusion power plant. On May 7, the *New York Times* optimistically reported on the characteristics of the new machine: "Britain Indicates Reactor Advance" read the headline. The following week, though, the paper planned a much less adulatory article: "H-Bomb Untamed, Britain Admits." The dream had come crashing down. Once again, the culprit was those damn false neutrons.

Even while the ZETA scientists were cracking open beers, toasting their first fusion reactions, Basil Rose, a physicist at Harwell, was consumed by skepticism. He was unconvinced that the ZETA neutrons were truly from thermonuclear fusion. While Cockcroft and others plotted their twenty-year path to fusion energy, Rose racked his brain for a way to allay his doubts. He simply had to come up with a way of proving that the neutrons were coming from fusion and not from a bizarre instability.

The method he came up with was analogous to what the Columbus scientists had done several years before: he would look at the symmetry of the system. Rose ran the ZETA machine twice, once in its normal operating mode and once with magnetic fields and currents reversed. If the neutrons were coming from a true thermonuclear reaction, the neutrons would have the same energies, no matter whether the machine was running normally or in reverse. The reaction should be symmetrical. It wasn't. The neutrons generated by normal ZETA had energies different from those produced by reverse ZETA. The neutrons weren't coming from thermonuclear fusion. They, too, were false neutrons.

As soon as Rose published his results in *Nature*—on June 14, a month after Britain revealed its plans for the ZETA II—it became obvious that scientists at the Harwell lab had deceived themselves. John Cockcroft immediately regretted his "90% certain" remark and assured the public that ZETA was a success even though it hadn't achieved thermonuclear fusion. "It is doing exactly the job we expected it would do and is functioning exactly the way we hoped it would," he sheepishly explained. However, the damage had been done.

Like Richter before them, the British had gotten burned for crying fusion. Driven by their optimism and goaded by their egotistical desire for glory, the ZETA scientists had humiliated themselves in front of the world. The stakes of fusion energy were so high—virtually unlimited power to the nation that controlled it—that scientists couldn't resist staking an early claim in achieving that lofty goal.

ZETA was a public relations disaster. For years, the cloud of ZETA hung over fusion scientists all over the world. In America, Project Sherwood physicists, despite their relief at not losing the race to the Brits, were despondent about what had happened. Fusion scientists were beginning to realize that fusion energy would be much more difficult to harness than they had thought. Fast pinches were not enough. There was no easy road to building a power plant with a magnetic bottle.

Even as scientists learned more about fusion, the dream of unlimited power seemed to slip further away. Another contender, though, was on the horizon, another way to confine a plasma and initiate a fusion reaction that would ignite another race for fusion energy.

CHAPTER 5

HEAT AND LIGHT

What glory beats in this idea:
Artificial suns on the earth,
Under controlled conditions.

—*RICHTER: THE OPERA*

Shortly after the ZETA defeat came a measure of victory. Unfortunately, nobody cared.

In 1958, not long after the British scientists at Harwell retracted their claim of generating thermonuclear fusion, American physicists finally put a tiny fusion reaction in a magnetic bottle. It was not a very large reaction, and starting the fusion consumed many times more energy than the reaction produced, but they had managed to initiate fusion in the laboratory. The neutrons they detected were from thermonuclear fusion.

The machine that did it was a pinch machine, not dissimilar to Columbus or the Perhapsatron, but its pinch was arranged slightly differently. Previous devices simply zapped an electric current down the length of the plasma. The new device, Scylla, ran a current *around* the circumference of the tube of plasma instead. It was just a variant on the existing pinch machines, but the change made a big difference. Scylla was able to heat deuterium to more than ten million degrees.

Scientists began to detect all the expected products of deuterium-deuterium fusion: protons, tritium nuclei, and neutrons. Tens of thousands of neutrons. With a few months of tinkering, physicists were getting roughly twenty million neutrons every time they ran the machine. It was a stunning success after so much failure.

It was just in time, too. The Second International Conference on the Peaceful Uses of Atomic Energy convened in September 1958. Though the ZETA fiasco was still on everybody's minds, the American display of fusion machines impressed visitors. Scylla made an appearance, along with the other magnetic bottles built by Project Sherwood. The Perhapsatron was there, as was Columbus. The early Stellarators also drew a crowd. Scylla should have been the star of the show, but the Scylla scientists weren't ready to make a formal announcement of their accomplishment. They were uncertain about whether they had truly achieved thermonuclear fusion and were well aware of the damage that a premature announcement could cause.

There were sly hints, of course. Los Alamos's Tuck gently implied that Scylla had succeeded where ZETA had failed, but he was much more cautious than the ZETA team had been. There was no press conference, just a scientific paper that stated, blandly, that Scylla "looks probable as a thermonuclear source." There were to be no adulatory headlines. Even when Tuck finally made a formal announcement—a year and a half later, in March 1960—it was to Congress, not to the press. "We are now prepared to stake our reputations that we have a thermonuclear reaction," he said. Scylla had done, for real, what ZETA had falsely claimed to do, but this time the world scarcely noticed.

The quest for unlimited fusion energy was in a dramatically different state than it had been a mere two years earlier. The public's attitude had changed: the ZETA affair and the growing concern about nuclear fallout had soured people's perception of fusion scientists. The scientists themselves were even growing pessimistic. Gone were the heady days of the 1950s when a working fusion reactor seemed to be just a few hundred thousand dollars away. Plagued by problems and instabilities, Project Sherwood seemed to be stalling. Congress, impatient with fusion scientists' broken promises, began to pull the plug on magnetic fusion research. Physicists raced to make some kind of discovery that would keep their quest for fusion energy alive. In 1958, the road ahead seemed dark.

In fact, there was a new reason for hope. The year brought a new and powerful idea into America's quest to tame fusion reactions—a novel Russian design that combined the advantages of the pinch and the Stellarator. It also saw the invention of a revolutionary device, the laser, that could bottle a tiny star in an entirely new way. The magnetic bottle was no longer the only game in town. A new set of hopefuls would soon sally forth to tame the power of the sun, only to be battered by the quest.

———

If there was one thing that scientists at the 1958 UN conference could agree about, it was that plasmas were proving very tough to control. In part, this was because plasmas were like nothing else scientists had encountered in nature. Plasmas behaved something like fluids, but unlike standard fluids, they interacted in extremely complex ways with magnetic and electric fields. Because of that electromagnetic component, understanding plasmas was becoming an entirely new discipline vastly more complicated than the hydrodynamics field that dealt with the behavior of ordinary fluids. Plasma physicists were charting new territory in a brand-new subject: magnetohydrodynamics. Even the simplest-sounding problems with a plasma turned out not to be simple at all.

What happens, for example, when you expose a plasma to an electric current, as in a pinch machine? The laws of electromagnetism say that electrical currents spawn magnetic fields, and magnetic fields spawn electrical currents. This means that an electrical current traveling down the plasma will generate magnetic fields that generate electrical currents that generate magnetic fields, and so forth—and all these effects change the motion of the particles in the plasma, forcing them toward the center of the cloud. This is why a current causes a pinch, confining the plasma and squeezing it into a tight thread. But this pinch has secondary effects, such as causing the thread to partition itself into little segments like a set of sausage links. The people who designed the pinch machines were immediately able to spot the pinch effect, but it took deeper thought to find the secondary effect of the sausage instability. The deeper the physicists looked into plasma dynamics, the more strange effects they saw—secondary and tertiary and beyond—most of which seemed to make the plasma unstable.

This feedback between electric fields and magnetic fields is just one of many effects that make plasmas hard to predict. Another has to do with the density of the plasma. Electric currents behave differently in plasmas of different densities and pressures. A current passing through a cloud of plasma alters the shape and density of the cloud—a pinch compresses the shape and increases the density—but this change alters the nature of the current passing through the cloud. This changes the shape and density of the cloud, which alters the current, which alters the shape and density of the cloud, and so on. Yet another issue had to do with the very makeup of the plasma. Scientists had been trying to ignore the fact that a plasma is not a nice, homogeneous substance made of a single kind of particle. A plasma is made of very heavy positively charged particles (the nuclei) and very light negatively charged particles (the electrons stripped from the atoms). These two kinds of particles have different properties and behave differently even when they are at the same temperatures, at the same pressures, and subjected to the same electromagnetic fields. Physicists discovered that when they tried to heat a plasma, unless they were very careful they would pour most of the energy into the light (and easy to accelerate) electrons, leaving the heavy nuclei cold, unheated, and slow. This was really bad news. The whole point of heating the hydrogen plasmas was to heat up the nuclei so that they were moving fast enough to fuse; hot electrons and cold nuclei were all but worthless. Unless scientists could compel the hot electrons to share their energy with the nuclei, there would be no hope of fusion. For all these reasons—and more—plasmas were very hard to work with. Even before a plasma gets hot and dense enough to ignite, it is a fiendishly complex brew.

The brew did not behave the way scientists expected it to. It seemed to have a mind of its own, thwarting all attempts to keep it under control. Pinch it or squeeze it or even try to keep it confined in a magnetic trap and it writhed around and ruffled itself in instability after instability. Physicists built bigger and more expensive machines to wrestle the instabilities into submission, but they were failing. As the machines started costing millions and tens of millions of dollars, the scientists were no closer to building a fusion reactor than before; they were just uncovering more and more subtle ways that the plasma fought their will.

Lyman Spitzer's Stellarator, for one, was mysteriously losing the particles in its plasmas. High-temperature particles move very quickly and are inherently hard to constrain. It was no surprise then that hot plasma particles in a crude magnetic bottle would rapidly spiral out of control and slam into the walls of the vessel, and the higher the temperature, the faster the particles were lost. On the other hand, increasing the magnetic field strength—strengthening the bars of the magnetic cage containing the plasma—should have rapidly brought this problem under control. That was the theory, anyhow. The researchers thought that if they doubled the magnetic field, they should cut the loss rate by a factor of four. If this theory was right, it would be fairly simple to get particle losses under control merely by cranking up the strength of the magnets surrounding the plasma. Relatively weak magnetic fields would suffice to confine even very hot plasmas.

Nature wasn't quite so kind to the Stellarator. As the scientists turned up the magnetic fields, they were surprised to discover that particles still zoomed out of control very quickly. The particle loss rates weren't dropping nearly as fast as the physicists' theories had led them to expect. Even with very powerful magnetic fields, the particles still spiraled out of control in a fraction of a second. Simply turning up the strength of the fields was not enough to bring the losses down to a reasonable level. The plasma was still out of control. The American physicists were beginning to despair.

Even the optimistic Spitzer gave up his dreams of a quick and cheap path to fusion energy with a Stellarator. In the early 1950s, he had thought that his small model-A and model-B Stellarators would lead quickly to a bigger, model-C machine that would serve "partly as a research facility, partly as a prototype or pilot plant for a full-scale power producing reactor." Spitzer was fairly certain that he would be within sight of a working fusion power plant by the end of the decade. Then, setback after setback sapped his optimism. By the late 1950s, he viewed the $24 million model-C Stellarator then under construction "entirely as a research facility, without any regard for problems of a prototype."* Spitzer no longer saw fusion energy as within his grasp; a pilot plant was many generations of machines away.

It was a difficult time for fusion physics. Even the successes of Sherwood, such as Scylla's first sighting of thermonuclear neutrons, were not showing a path to a working reactor. Making matters worse, the Atomic Energy Commission's budget, which had skyrocketed through the 1950s, stopped growing, and the fusion research budget itself began to pinch.

These difficulties bred a measure of hope for magnetic fusion. Perhaps because the goal of a fusion reactor was so far out of reach, all the nations working on fusion energy decided to share their knowledge. The stakes had been lowered; there was no obvious path leading to limitless energy, so there was no harm in international collaboration. At the 1958 UN conference, the shroud of secrecy finally lifted from the fusion reactor programs around the world. Not only did American and British physicists have permission to lecture about the work they had done over the past decade, so, too, did their Russian counterparts. And behind the Iron Curtain, Soviet physicists had been doing some extraordinarily good work. The West soon learned of an idea that came from Russia's version of Edward Teller: Andrei Sakharov.

Sakharov was a little more than a decade younger than Teller, so he was still a student when World War II erupted. He built a wartime reputation by working on conventional, not nuclear, munitions. He came up with a clever method to use electric and magnetic fields to detect

---

* Even the expensive model-C Stellarator, whose price had swollen 50 percent since its proposal, wasn't much of an improvement over the earlier models. When it first came on line in the early 1960s, it was only better than its predecessors by virtue of its larger physical size, which meant that it took longer for a particle to stray far enough to strike a wall.

defective armor-piercing shells, a vast improvement over the backbreaking work of snapping random shells in half to see whether they were properly manufactured. As the war was ending, Sakharov returned to school to get a graduate degree in physics, thinking he had escaped his weapon-engineering days. But on August 7, 1945, he was drawn back to military work.

On his way to the local bakery, Sakharov happened to glance at a newspaper. It told of the destruction of Hiroshima. "I was so stunned," he wrote, "that my legs practically gave way.... Something new and awesome had entered our lives, a product of the greatest of the sciences, of the discipline I revered." The cloud of the atom bomb began to mushroom over his studies. As Sakharov tried to concentrate on theoretical physics, those mysterious secret cities began to spring up across the nation. His mentor, Igor Tamm, was secretly getting involved in Russia's nuclear program. By 1948, Sakharov had been drawn into a project to design fusion weapons (the atom bomb problem having already been worked out, in part, thanks to the spying of Klaus Fuchs).

Sakharov immediately came up with his "first idea," a design for a thermonuclear weapon. This design, the sloika bomb, was almost identical to the layered Alarm Clock design that Teller discarded as impractical in 1946. Though the sloika had the same problems as the Alarm Clock—megaton-size weapons would be too large to be practical—it offered a quick path to building a fusion device. Sakharov's first idea impressed the Kremlin, which then whisked him away to one of the secret cities. To his dying day, Sakharov only referred to the laboratory as "the Installation," and for a time, its mere existence was one of the most closely guarded secrets of the Soviet Union. The Installation was an entire town, code-named Arzamas-16, built for the purpose of designing nuclear weapons.

Sakharov's intellectual trajectory was an eerie mirror image of Teller's, always delayed by a few years.* Teller came up with the imprac-

---

* Sakharov saw even deeper similarities between himself and Teller, going far beyond the physics and into their motivations for participating in the arms race. "One had only to substitute 'USSR' for 'USA,' 'peace and national security'

tical Alarm Clock configuration for the hydrogen bomb in 1946; Sakharov hit on his sloika in 1949. In 1946, Teller proposed boosting the yield of a fission bomb by injecting a tiny dollop of fusion fuel (the idea tested in Greenhouse Item). Boosting atom bombs was Sakharov's "second idea," which came soon after his first. In 1949, Ulam and Teller solved the problem of igniting a fusion reaction by separating the primary fission device from the secondary fusion one; Sakharov and his colleagues came to the same solution—Sakharov's "third idea"—in 1953.

Not everything, though, occurred to the Americans first, especially when it came to fusion reactors. In 1950, Sakharov was hard at work trying to figure out how to build a hydrogen bomb—an uncontrolled fusion reaction—when he began to ponder whether the reaction could be controlled. Like his American counterparts, he came up with a scheme using magnetic fields, but his idea was slightly different from the ones that would guide Project Sherwood. Sakharov's device was neither a Stellarator nor a pinch machine. It was somewhere in between. It was a novel design, one that combined some advantages of a pinch machine with those of a Stellarator. It was just what scientists were looking for.

In a pinch machine, the plasma confines itself. The pinch begins by inducing a current of some sort in the plasma, forcing it to contract and to heat up. The Perhapsatron, ZETA, and Columbus all did this by running a current down the length of the plasma, while Scylla ran an electrical current around the circumference of the tube of plasma. In both cases, though, there is a current inside the plasma; this current induces a magnetic field, which squashes the plasma. Pinch machines

---

for 'defense against the communist menace,'" Sakharov later wrote. This was something of an exaggeration, though. Sakharov did not share Teller's almost monomaniacal hatred of the enemy. Because of this, their paths would soon dramatically diverge, beginning with the debate over fallout. Sakharov would become a firm opponent of atmospheric nuclear testing and a fearless proponent of a test ban and international cooperation. While Teller was honored by his country and reviled by the Nobel committee for his actions, Sakharov went into internal exile in the Soviet Union and received the 1975 Nobel Peace Prize for his.

were successful at making a plasma very dense and hot, even inducing a bit of fusion, but physicists couldn't keep those conditions going for very long. The pinch was extremely short-lived. Once the current disappeared, so did the confinement of the plasma. Given the enormous energy it took to set up a pinch, and how little energy was generated by the brief fusion reaction, a pinch machine could not ultimately become a working reactor.

In a Stellarator, on the other hand, the plasma is confined from the outside. The machine uses carefully arranged electromagnets to generate intricate magnetic fields that bottle up the cloud. In theory, these fields would allow a Stellarator to confine the plasma for a relatively long time. Unlike a crush-and-release machine, a Stellarator attempted to maintain its hold on the plasma, keeping a fairly stable cloud. But scientists were having difficulty not only confining the plasma but also heating and compressing the cloud. It was much harder to warm and squash plasmas in a Stellarator than it was in a pinch machine.

The choice was bleak for American scientists in the late 1950s and early 1960s. They could get high temperatures and densities for a short time or lower temperatures and densities for a longer time, but not both. Yet scientists really needed a magnetic bottle that could heat the plasma to tens of millions of degrees, keep it very dense, and hold it for a relatively long time. The heating and density would ensure that the fusion reaction took place, while the confinement would ensure that the plasma reacted long enough to generate a significant amount of energy. Only then could physicists hope to turn such a magnetic bottle into a fusion reactor.

Sakharov's scheme appeared to provide an answer. His bottle looked little different from the ones the Americans and British were proposing. It was donut shaped—toroidal—and used coils of wire to induce magnetic fields, earning it the cumbersome name *toroidalnaya kamera ee magnitnaya katushka* (toroidal chamber with magnetic coil). It was called the *tokamak* for short. But the tokamak was a bottle with a difference. Whereas the Stellarator used external magnetic fields to contain the plasma and the pinch machines used internal electric currents to squash it, the tokamak did both.

The tokamak has multiple sets of coils. One group of coils sets up a magnetic field that constrains and stiffens the plasma; it's an external magnetic bottle, somewhat similar to the Stellarator's, although not quite as sturdy. What gives the tokamak an extra bit of oomph is another set of coils that pinches the plasma. When scientists send a current through those coils, it induces a corresponding pinching current in the plasma circulating in the torus. This one-two punch of the external magnetic fields and internal current gave scientists a tool that, they hoped, would keep a hot, dense plasma stable for a long time.

Of course, the tokamak design had drawbacks as well. Unlike the Stellarator, which doesn't require a plasma current at all, a tokamak absolutely needs one; its external magnetic bottle can't by itself contain the plasma cloud for very long. But reducing this plasma current adds layers of complexity to the plasma, making it more unpredictable.

In some sense, the tokamak is something like a bicycle. Just as a bicycle is not stable until it is going relatively quickly, a tokamak's plasma is not stable until the plasma current is up and running. The Stellarator is more like a tricycle. Just as a tricycle can be stationary, or can move forward or backward without any threat to its basic stability, a Stellarator can either have no plasma current or have one in either direction and still, theoretically, be stable.

Unfortunately, the current in a tokamak's plasma is just one more thing that can fail. If an instability causes the current to drop momentarily, things get very bad very quickly. The plasma suddenly loses its pinch and explodes in all directions. This event is called a *disruption*, and it can be extraordinarily violent. It can even damage the machine. (One disruption at a modern British tokamak made the whole thing, all 120 tons of it, jump a centimeter into the air.) However, the disadvantages of the tokamak soon seemed small compared to the advantages of the design.

Sakharov was too busy working on nuclear weapons to spend a lot of effort on fusion reactors. But other Russian scientists, particularly the physicist Lev Artsimovich, took Sakharov's design and put it to the test. By the mid-1960s, he was reporting spectacular results. His tokamak was confining a plasma at a given temperature and density ten times longer than could any other machine. Though confinement times were still on the order of milliseconds, Artsimovich's results, if they were to be believed, indicated that Sakharov's tokamak was blowing its competition away.

When Spitzer and the Americans first heard the Russian claims, they were skeptical, in part owing to American arrogance. The Stellarator was performing quite poorly, so they concluded that the problems they were encountering were likely due to a universal problem with magnetic confinement. If they weren't succeeding, nobody was. The Americans were dubious that the Russians could do much better with their tokamak. Furthermore, Artsimovich's measurements of the temperature of the plasma were rather crude. American scientists were relatively quick to disbelieve them. In the mid-1960s, Spitzer's skepticism led him to conclude that tokamak performance was roughly the same as the Stellarator's—underwhelming.

This conclusion was bad news for American fusion research. The enthusiasm of the 1950s had brought a downpour of funding from Congress. Since Project Sherwood's inception, its budget had skyrocketed from almost nothing to nearly $30 million a year by the time of the 1958 UN conference. As the Stellarator began to choke, losing its plasma rapidly, a skeptical Congress began to wonder whether fusion reactors were possible at all, much less economically feasible. It didn't help that the scientists, in their optimism, had consistently oversold their machines. They had promised Congress they would be building prototype reactors by the early 1960s, and the machines were nowhere near that stage. And in October 1957, the Russian surprise launch of Sputnik gave the Earth an artificial moon—and it gave Congress another Cold War scientific competition requiring truckloads of taxpayer money. The space race had officially begun. Fusion energy was no longer in the spotlight, and its budget stagnated, then dwindled.

The tokamak had to come to the rescue, but it would be several years before American and British scientists would accept that the Russian achievements were real. It was not for lack of data; Artsimovich continued presenting better and better results—dense plasmas heated to tens of millions of degrees and confined for handfuls of milliseconds. The tokamak results were still far from those needed for a realistic source of fusion energy, but they were certainly an order of magnitude better than anyone else's. The work was getting harder to dismiss, but detractors still argued that the Russian temperature measurements were inaccurate. To settle the matter, in 1969 a British team visited Artsimovich's lab in Moscow. They came armed with a sensitive instrument that could measure the temperature of a ten-million-degree plasma.* At the heart of the instrument was a device that would change the face of fusion research, and not just because it confirmed the Russian claims. It would provide a new way of bottling a plasma without the use of magnets. The device was the laser.

Depending on whom you ask, the laser was invented at Columbia University in the late 1950s or at Hughes Research Laboratories in 1960. (There were competing claims and a patent battle.) But there's no doubt that in 1960 a short paper in *Nature* gave the physics community a powerful new tool.

A laser is a device that produces an unusual beam of light. Even to the uninitiated, it is obvious that laser light is different from, say, the light that comes from a flashlight. If you shine a flashlight at a distant wall, you will see that it makes a large, circular spot. If you shine a laser pointer at the same wall, it makes only a tiny dot, barely larger than the hole from which the laser beam emerged. Laser light stays together in a tight beam rather than spreading out into a diffuse cone. A laser beam also consists of light that is a single, intense color,† unlike a flashlight

---

* That this could be done at all baffled the people who funded the experiments. In a debate in Britain's House of Lords, one peer asked another how scientists could measure temperatures of tens and hundreds of millions of degrees. The response? "I expect that they use a very long thermometer."

† Physicists use several terms when referring to the color of light: *color, energy, frequency,* and *wavelength* all essentially refer to the same thing. Visible light comes

beam, which is made of a whole bunch of colors mixed together and appears white.

There are many methods of generating light. If you heat something high enough, it begins to glow. When a substance is energetic enough, it emits visible light. (This is how an incandescent lightbulb works; the filament in the bulb is simply heated to a very high temperature.) It is a law of nature: the hotter an object is, the more light waves it emits. Or, if you prefer, you can think of the emissions as light particles rather than light waves. The laws of quantum theory say that light has both a particle-like and a wave-like nature, so physicists use whichever description is most suitable for the behavior they are attempting to describe.

A particle of light—a photon—can interact with matter in a number of different ways. It can strike an atom and give it a kick. It can make the atom rotate or move in other manners. If the photon is just the right color, the atom can absorb it. Absorbing a photon "excites" the atom, packing it full of the energy that once resided in the light particle. This excited atom will soon disgorge the photon, emitting a light particle of precisely the same color and relaxing from its excited state.

In 1917, Albert Einstein made a curious prediction about excited atoms. Such an atom is quivering with energy, looking for an excuse to spit out the photon it has absorbed. Einstein's calculations showed that if a photon of the right color happens by—one precisely the same color as the one absorbed by the atom—then the atom will immediately disgorge a photon. This photon not only will be precisely the same color as the passerby but will also move with it in lockstep. The two photons will behave almost as a single object. This phenomenon is known as

---

in a variety of colors: a whole spectrum stretching from red on through orange, yellow, green, blue, and indigo, to violet. It turns out that these colors correspond to light of higher and higher energy. Red has the lowest energy and violet has the highest. Frequency is proportional to energy, so red has the lowest frequency and violet the highest. Wavelength is inversely proportional to frequency, so red has the longest wavelength and violet has the shortest. It can get confusing, but the terms are all trying to describe the same phenomenon in different ways.

stimulated emission, and it is the mechanism the laser uses to produce its beam of light.*

Imagine that you have a hunk of material—a whole lot of atoms—that you want to turn into a laser. The first step is to excite all the atoms. You do this by "pumping" the material full of energy. It doesn't matter how. Some lasers pump a material with electricity. Some lasers do it with light, and some do it with chemical reactions. It might even be possible to use nuclear bombs to pump atoms into an excited state.† Once the atoms in the material are excited, they are primed to get rid of their energy—they want to emit light of a particular color.

This is where the clever part happens. Send a photon of that specific color into the material. The photon encounters an excited atom, which then disgorges a second photon of the same color through stimulated emission. These two photons move in lockstep. They encounter another excited atom, which emits another photon of the same color: three photons now in lockstep. Another excited atom, another photon: four photons, all the same color, all moving in precisely the same way. As the photons move through the material, they encounter more and more excited atoms, which emit more and more photons. The beam snowballs, growing bigger as it travels through the material. By the time it finally emerges, the beam consists of an enormous collection of light particles. It is an intense beam, and all the photons have the exact same color and are moving in lockstep, almost like one enormous particle of light. This is the secret to the laser's power. It is why the photons in a laser beam don't zoom out in different directions and have all sorts of colors as a flashlight's do. The photons in an ordinary beam of light are like are an unruly mob; the photons in a laser beam are an army marching together with a single mind.

The laser's unusual properties make it an incredible scientific tool.

---

* Hence the name "laser," which is an acronym for light amplification by stimulated emission of radiation.
† Thank Edward Teller for this last proposal. In the 1980s, he pushed hard to design a bomb-powered laser that could shoot down enemy missiles; the concept was a big part of Ronald Reagan's "Star Wars" plan.

The tight beam allows it to travel great distances—to the moon and back, even—without scattering and dissipating too much. Because the beam is made of photons of the exact same color, it provides a great way to measure very, very hot temperatures.

Shine a laser at a plasma. The photons in the beam will begin with the exact same color. But as the photons strike the fast-moving particles in the plasma, the plasma gives the photons a kick, adding a bit of energy to them, shortening their wavelengths and making them slightly bluer. By looking at the color of a laser beam after it hits a plasma, scientists can calculate the energies of the particles in the plasma, which, in turn, reveals the temperature.

When the British scientists shined a laser beam at Artsimovich's tokamak plasma, they saw that the Russians were not exaggerating. Their plasma was tens of millions of degrees, dense, and relatively well confined. The tokamak *was* performing much better than the other forms of magnetic bottles. This was wonderful news for the fusion community in the West, even though the Russians, rather than the Americans or the British, had done it. (One Atomic Energy Commission worker reportedly danced on a table when he heard the news.) Sakharov's invention showed a way to bypass the troubles of the pinch machines and the Stellarators. Practically overnight, plasma physicists across the world scrapped their old devices and built tokamaks. Even Spitzer succumbed to tokamania. By January 1970, the model-C Stellarator was scrap metal. In its place, a mere four months later, a tokamak sprang up. The fusion community had been pumped full of energy once more.

The laser measurements of Artsimovich's plasma sparked a fusion energy renaissance in the United States. But the laser was about to change the landscape even more dramatically by providing an alternative to the magnetic bottle.

Lasers produce particularly intense and yet easily controlled light beams. You can point a laser with great precision and make it dump an enormous amount of energy in a very tiny space. To Andrei Sakharov, this suggested that laser beams could be used to heat and contain a plasma of hydrogen. If it worked, laser fusion would be an even more straightforward method than that using magnets. One could simply shine laser light on a pellet of deuterium fuel from all directions: the beams would heat and compress the pellet, creating a tiny fusion reaction—a miniature sun girdled on all sides by light. The plasma would be compressed not by magnetic fields but by particles of light (or by atoms that had been heated by the beams of light).* This was the birth of *inertial confinement fusion.*

The Americans, too, immediately saw the potential of lasers for inducing fusion. At Teller's Livermore laboratory, physicists like Ray Kidder, John Nuckolls, and Stirling Colgate set to work designing laser fusion schemes soon after the first laser was built.† Their calculations seemed to show not only that laser fusion was possible, but also that it might be relatively easy to achieve breakeven. Livermore scientists began building laser bottles intended to ignite and contain fusing plasma.‡

The first big one, built in 1974, was known as Janus. Two-faced like the god it was named after, Janus had two laser beams that shot at a tiny pellet of deuterium and tritium from opposite directions. It was more a test of the laser system than a concerted attempt to initiate fusion reactions. A true laser-based bottle would require laser beams to hit the target from all sides at once to fully confine it, but Janus's lasers only struck from two sides, allowing the plasma to squirt out in various directions. Nevertheless, the Livermore scientists were soon detecting tens of thousands of neutrons coming from the pellet. They had achieved thermonuclear fusion, even though it was on a tiny scale. It was a success, but it was not the first.

The Russians and French had already detected neutrons from pellets hit by lasers, but the American press, skeptical of the foreigners'

---

* In practice, this usually means the latter rather than the former. In most modern inertial confinement fusion experiments, light heats up the outermost layers of a target capsule, causing them to evaporate. This pushes the rest of the capsule inward and ignites the fuel.

† Nuckolls had been involved in Project Plowshare. In the late 1950s, he designed a fusion power plant that heated steam by means of hydrogen bombs.

‡ Other labs, such as Los Alamos, had inertial confinement fusion programs, too. Livermore, however, came to dominate.

claims, did not give them much attention. The press did have a field day, though, with the curious tale of a rogue company—KMS Industries, Inc.—that had built its own laser system. By May 1974, KMS, named after its physicist founder and president, Keeve M. Siegel, reported that it was producing neutrons from laser fusion.

Within two weeks, the story was plastered all over the newspapers. The *New York Times* touted KMS's achievement as "a significant step toward the long-range goal of nuclear fusion as a source of almost limitless energy." The Atomic Energy Commission was less thrilled, because a private firm was doing an end run around the government. If the KMS claims were true, an AEC statement read, it would be "a small but significant initial step toward the achievement of fusion power." Siegel was making the AEC look bad—and fusion energy look good.

Not only was Siegel using lasers to ignite fusion, but he was doing it as the head of a private company, not as a scientist in a government laboratory. The public took this as a sign that private industry was embracing fusion reactors as a viable source of energy. Siegel, the entrepreneur, exuded confidence in public. He was sure, he said, that he could turn lasers into "efficient fusion power" within "the next few years." After false starts and two decades of struggle with magnetic bottles, the era of fusion finally seemed at hand.

The timing could scarcely have been better. The United States was just getting through its first oil crisis. Because of American support for Israel during the 1973 Yom Kippur War, the Arab members of the Organization of the Petroleum Exporting Countries (OPEC) cut off oil supplies to the U.S. Gas prices skyrocketed. It was becoming painfully clear that the country had to find another source of energy—anything other than petroleum—if it was to avoid being held hostage to OPEC's interests. It was scarcely two months after the embargo was lifted that a jittery nation learned about Siegel and KMS. It seemed that fusion would be the way to get out from under OPEC's thumb. The dream of unlimited power once more beckoned. Fusion energy seemed possible again, and it was more important than ever.

Congress immediately seized upon it and started pouring money into fusion research. Laser fusion saw a dramatic increase in funding, growing from almost nothing to $200 million per year by decade's end.* Livermore and some other laboratories around the country, particularly those at Los Alamos and at the University of Rochester in New York, began to plan massive laser projects with an eye toward creating a viable fusion reactor. Magnetic fusion, too, benefited from the renewed interest in fusion energy. After stagnating for a decade at around $30 million per year, magnetic fusion budgets doubled and doubled and doubled again. In 1975, more than $100 million went to magnetic fusion; by 1977, more than $300 million; and by 1982, almost $400 million.

Siegel's 1974 announcement helped ignite public enthusiasm (and governmental largesse) for fusion research, but his story had a tragic ending. In 1975, he keeled over while testifying about his work in front of Congress. Though he was rushed to the nearby George Washington University Hospital, he died shortly thereafter, the victim of a stroke. He was fifty-two years old. Siegel didn't survive to benefit from the surge of optimism he generated. He also didn't survive to see the worsening problems laser fusion scientists faced as their lasers grew more powerful.

Livermore's Janus was already in 1975 suffering from a major snag. Its lasers were extremely powerful for their day, pouring an unprecedented amount of laser light into very tiny spaces. Livermore's scientists managed to get this level of power by taking enormous slabs of glass made of neodymium and silicon and exciting them with a flash lamp. This glass was the heart of Livermore's laser. The slabs were what produced an enormous number of infrared photons in lockstep. The resulting beam exited the glass and was bounced around, guided by lenses and mirrors to the target chamber. However, the beam was so intense that it would heat whatever material it touched. This heat

---

* Congress also took the opportunity to reorganize its entire portfolio of energy research. In 1974, it eliminated the Atomic Energy Commission and created the Energy Research and Development Administration to take on many of its functions. Just a few years later, ERDA, together with other federal agencies, would become the Department of Energy.

changed the properties of lenses, mirrors, and even the air itself. When heat changes the properties of a lens or a mirror, it alters the way the device focuses the beam. These little changes in focus would start creating imperfections in the beam, such as hot and cold spots. These could be disastrous. The hot spots in the beam would pit lenses, destroying them in a tiny fraction of a second. Every time they fired the Janus laser, the machine tore itself to shreds.

Luckily, the Livermore scientists were already working toward a fix. Their next-generation fusion machine, Argus, used a clever technique to eliminate those troublesome hot spots. By shooting the beam down a long tube and carefully removing everything but the light at the very center of the beam, the scientists would be assured of getting light that was uniform and pure—and free of hot spots. This meant that the laser had to be housed in a very large building to accommodate the tubes, which were more than a hundred feet long. In addition, since they were tossing out some of the beam because of its imperfections, they were sacrificing some of the laser's power. This was a minor inconvenience; the technique worked, and the hot spots disappeared for the time being.

More serious was the problem with electrons. Magnetic fusion researchers had trouble heating the plasma evenly; the lightweight electrons would get hot faster than the heavyweight nuclei, making for a very messy plasma soup. This problem was worse with lasers: light that is shined on a hunk of matter tends to heat the electrons first. This was a huge issue. The electrons in a laser target would get so hot that the target would explode before the nuclei got warmed up. Hot electrons and cold nuclei were no good for fusion—it was the nuclei that scientists really wanted to heat up.

For technical reasons, the bluer the laser beam, the smaller this effect. So the Livermore scientists shined the laser light through crystals that would make the infrared beam green or even ultraviolet.* The color conversion worked well to reduce the heating of the electrons, but the process was inefficient. The beam lost some of its energy be-

---

* Technically, doubling or tripling the frequency of the light.

cause of the color change. It also made the laser more expensive, as big, high-quality color-change crystals were not cheap. Nevertheless, the results—and the number of neutrons—from Argus led Livermore's physicists to push for a full-size machine, Shiva, that would use twenty beams to zap a pellet of deuterium from all directions. It would ignite the pellet, creating a fusion reaction that would generate as much energy as the laser poured in. Or so the scientists hoped. They were wrong by a factor of ten thousand. Laser fusion scientists, like the magnetic fusion advocates that preceded them, were about to come face-to-face with a nasty instability—one so fundamental that you often encounter it in your kitchen.

It is hard to imagine an instability in the kitchen, but ask yourself the following question: When you invert a glass of water, why doesn't the water stay in the glass? This seems like a silly thing to ask: gravity pulls the water down and onto the floor. But if you look a little more deeply, the answer is not quite so obvious. Atmospheric pressure makes the question more complicated than you might expect.

Every surface that is exposed to air is under pressure. The very weight of the atmosphere is squashing us from all directions. Every square inch of our skin is subjected to 14.7 pounds of pressure from the air pushing against us. We don't notice it because our bodies are used to it, but this is an enormous force, easily enough to crush a steel can under the right conditions. It is also more than enough to support a glassful of water and prevent the liquid from falling to the ground. Try it yourself (over a sink, of course). Fill a glass to the rim with water. Hold a smooth, rigid piece of cardboard over the mouth of the glass and invert the whole thing. Gently let go of the cardboard. If you do it carefully enough, you will see that the water stays in the glass. The cardboard isn't holding the water in. It's not stuck tightly to the glass; even a gentle touch will dislodge the cardboard and cause the water to run out. And the water isn't miraculously defying gravity. It is being supported by air pressure. The atmosphere's upward push of 14.7 pounds per square inch is much, much stronger than the three or four ounces per square inch downward push of the water in the glass. When the two pressures go head to head, the upward push of the atmosphere wins and the water

RAYLEIGH-TAYLOR INSTABILITY IN A GLASS OF WATER: Invert a glass quickly and little ripples on the surface of the water will grow, becoming large blobs. The blobs break off and the water rains down out of the glass.

stays put. Believe it or not, the forces are so mismatched that you would need an enormously tall glass of water—about thirty feet high—if you wanted the downward-pushing weight of the water to equal the upward-pushing atmospheric pressure. With such vastly mismatched forces, the question seems a lot less stupid: Why doesn't water stay in a glass when you turn it upside down?

The water falls out because of an effect known as the Rayleigh-Taylor instability. Whenever a not-very-dense fluid (like air) pushes on a denser fluid (like water), it is an inherently unstable situation. If the interface between the two fluids has any imperfections—any bumps or divots—then those imperfections immediately get bigger and bigger.

An inverted glass of water, no matter how carefully it is inverted, has a few crests and troughs on the surface of the liquid. In a tiny fraction of a second, the crests grow, becoming enormous tendrils of water drooping down from the surface; the troughs also grow, and large fingers of air prod deep into the glass. The tendrils break, the fingers bubble off, and the entire glass of water rains down onto the floor. This is the Rayleigh-Taylor instability in action. Even though the air exerts an enormous amount of pressure on the water, the less-dense air is unable to keep the denser water contained in the glass because of these growing tendrils and fingers. Get rid of those instabilities and the air can keep the water contained. (The cardboard is not susceptible to Rayleigh-Taylor instabilities because it is a solid, so the air-pushing-on-cardboard-pushing-on-water system is stable.) But if Rayleigh-Taylor instabilities are present, then they will wreak havoc on your attempt to keep the denser fluid contained where you want it.

Laser fusion is the equivalent of keeping water trapped in an upside-down glass. As you compress a pellet of deuterium, it becomes denser and denser. Long before you get it hot and dense enough to fuse, it will be much denser than whatever substance you are using to compress it, whether it is particles of light or a collection of hot atoms. You are using a less-dense substance to squash and contain a much denser one, and that means you will get Rayleigh-Taylor instabilities. Any tiny imperfections on the interface between the plasma and the stuff that is pushing on the plasma will immediately grow. Even an almost perfectly round sphere of deuterium will quickly become distorted, squirting tendrils in all directions. Just as this ruins any attempt to keep water in an inverted glass by means of air pressure, it seriously damages a machine's ability to compress and contain a plasma

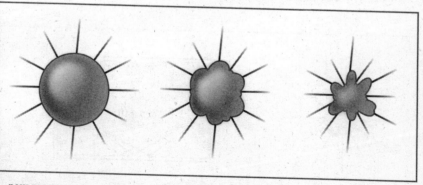

RAYLEIGH-TAYLOR INSTABILITY IN INERTIAL CONFINEMENT FUSION: Use lasers or particles to bombard a pellet of fuel and small imperfections on the surface of the pellet quickly become large fingers that cool the fuel and prevent it from fusing properly.

by means of light. The only way around this was to make sure there were almost no imperfections. The target had to be perfectly smooth, and the compressing lasers had to illuminate the target completely uniformly, without any hot or cold spots that would lead to ever-growing Rayleigh-Taylor tendrils.

It was almost as if the laser scientists were trying to invert a glass so carefully that the surface of the water inside wouldn't ripple at all. This is an extraordinarily difficult task. Even the twenty-armed Shiva machine, heating the plasma from twenty different directions at once, wasn't uniform enough to keep the Rayleigh-Taylor instabilities in check. The twenty pinpricks of laser light were far enough apart from one another that they would create hot spots in the target rather than heating it uniformly. The pellet would compress, getting hot and dense enough to induce a little bit of fusion, but before the reaction really got going, the Rayleigh-Taylor instability would take over. Tendrils would form. Instead of getting denser and hotter, the deuterium would squirt out.

The Livermore scientists tried everything they could to get the Rayleigh-Taylor problem under control. One method mimicked the Teller-Ulam design for the hydrogen bomb. Instead of using the lasers to push directly onto a dollop of deuterium, the new method did it indirectly. The pellet was ensconced at the center of a hollow cylinder

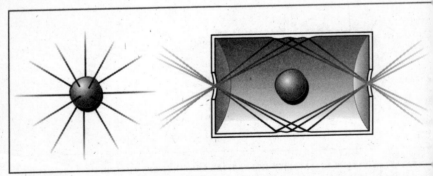

DIRECT DRIVE VERSUS INDIRECT DRIVE: In direct drive (left), laser beams shine directly on a pellet of fuel. Indirect drive (right), on the other hand, has laser light shining on a hohlraum, which evaporates and shines x-rays on the pellet.

known as a *hohlraum*. Instead of striking the pellet, the lasers struck the insides of the hohlraum. The hohlraum then radiated x-rays toward the pellet. This setup is known as indirect drive, and it helped ameliorate the problems with the instabilities.

But it didn't do enough. Shiva, which had cost $25 million to build, only performed a fraction as well as its designers had hoped. It didn't come close to producing as much energy from fusion as it took to run the lasers. Reaching breakeven was a much harder task than expected. The answer seemed within reach, though: just build a bigger Shiva, one with ten times the power, and ten times the price. By the beginning of the 1980s, Livermore was building a $200 million laser named Nova. Researchers there were confident Nova would finally take them to the promised land—igniting fusion fuel, producing more energy than it consumed. Once more, fusion scientists were about to have their faith severely tested.

The science of inertial confinement fusion was following the same trajectory as that of magnetic fusion. Early optimism in the 1950s led scientists to believe that plasmas could be confined and induced to fuse relatively easily. Cheap, million-dollar machines, they thought, would be able to do the job. But the plasma always seemed to wriggle out of control. Instability after instability made the magnetic bottles leak, and million-dollar machines turned into ten-million-dollar and hundred-million-dollar machines. Laser fusion began with similar optimism. Livermore's scientists thought their first few lasers could get more energy out than they put in. But instabilities like Rayleigh-Taylor allowed the plasma to escape its confinement. Million-dollar lasers grew bigger and more expensive. Soon, laser fusion machines were as expensive as their magnetic counterparts.

Even today, decades later, these two approaches—magnetic fusion and inertial confinement fusion—remain the ways that most scientists are trying to bottle up a tiny sun. But both methods are extremely expensive, and both are plagued with instabilities that threaten to destroy the dream of unlimited fusion energy. Shiva's failure occurred two decades after Homi J. Bhabha predicted that fusion power plants were twenty years away. Yet in the 1970s, and even into the 1980s, fusion scientists spoke of power plants as being thirty years away. After decades of research, the goal of fusion energy had become ten years more distant.

As fusion scientists built ever-bigger tokamaks and lasers for tens and hundreds of millions of dollars, outsiders began to wonder whether there was another cheaper, easier path to fusion energy. The stage was set for the biggest scientific debacle of modern times: cold fusion.

CHAPTER 6

THE COLD SHOULDER

We are also human, and we need miracles, and hope they
exist.

—LEONID PONOMAREV, FUSION SCIENTIST

An intricately crafted glass mushroom on a metal pedestal,
the two-foot-tall machine dominated the room—even when
it wasn't running. But when the operator twisted a dial and
brought the BioCharger to life, everybody stopped to look.
The helical glass coil at the top of the mushroom glowed red, and the
whole machine throbbed with electricity. Tubes running up and down
the mushroom's stalk fluoresced with blue and red light. It crackled
ominously as strands of violet lightning shimmied down the sides and
dissipated into the air. As the crowd stood transfixed, the smiling operator turned the dial back and the machine died abruptly. The smell
of ozone lingered in the air.

The BioCharger is a device that supposedly transmits healing energy directly into your body. Its inventor swears that the machine will
help cure your thrush, fatigue, diarrhea, night sweats, frequent urina-
tion, colds, unrefreshed sleep, and almost anything else that ails you.
The machine wouldn't ordinarily be allowed anywhere near a scientific
conference, but the BioCharger wasn't out of place at this one. Neither

was the device to test how much mercury was in your mouth to help diagnose the causes of your diseases, nor the presentation that discussed the "energy chair": an ordinary white plastic lawn chair with a generator underneath. ("We used to call it the electric chair, but figured we had to change the name," the presenter said.) The chair supposedly leaves you refreshed and energized after sitting in it. At an ordinary scientific gathering, such claims would be laughed out of the building. But the Second International Conference on Future Energy was no ordinary scientific conference.

Held in September 2006 on the outskirts of Washington, DC, the Conference on Future Energy was a celebration of sorts. Its convener, Thomas Valone, had recently won a long legal battle with his employer, the U.S. Patent and Trademark Office. Valone was a patent examiner who had, in his view, been fired for his belief in cold fusion. A year after being reinstated in his job (with back pay), Valone called a gathering of researchers together to, once again, explore the future of energy: a future that includes cold fusion.

Cold fusion had burst upon the world nearly two decades earlier and had long since been discredited by the mainstream scientific community. Yet today it still has a strong following, a core of true believers who think it will help humanity unleash unlimited power from fusing atoms. Plenty of reporters, government officials, and even scientists remain under its spell. The dream of unlimited energy through cold fusion is so powerful that for almost twenty years the faithful have been willing to risk ridicule and isolation to follow it.

———

The biggest scientific scandal of the twentieth century began on March 23, 1989. Two chemists at the University of Utah, Martin Fleischmann and Stanley Pons, told the world that they had tamed the power of fusion energy at room temperature, bottling up a miniature star in a little hunk of metal. The university's press release was full of enthusiasm:

SALT LAKE CITY — Two scientists have successfully created a sustained nuclear fusion reaction at room temperature in a chemistry laboratory at the University of Utah. The breakthrough means the world may someday rely on fusion for a clean, virtually inexhaustible source of energy.

At the press conference, the president of the university, Chase Peterson, pronounced that the scientists' discovery "ranks right up there with fire, with cultivation of plants, and with electricity." Yet such a monumental achievement came in a small and homely package. When Pons and Fleischmann displayed slides of their "reactor," goggle-eyed reporters were stunned. The apparatus was little more than a small glass beaker mounted in a dishpan. The claim rattled around the globe in a matter of hours, astonishing physicists and igniting a tremendous controversy. Over the next few weeks, skeptics expressed graver and graver doubts about the Utah chemists' claims, but other laboratories seemed to confirm their findings: in Utah, Georgia, Texas, Italy, Hungary, the Soviet Union, and India. The story of cold fusion quickly became a knotty mess that, decades later, has yet to be untangled.

Most physicists were immediately skeptical of the chemists' claim, and it is easy to understand why. Pons and Fleischmann were stating that they had caused deuterium nuclei to fuse in a little jar at room temperature. This seemed to contradict everything that physicists knew about nuclear fusion. Because the positively charged deuterium nuclei must slam into each other at very high speeds to fuse, it means that fusion tends to occur only when the deuterium is at a very high temperature and high pressure. This, of course, was why fusion scientists were spending hundreds of millions of dollars on lasers and magnets to heat and confine deuterium plasmas.

Pons and Fleischmann's setup was supposedly making an end run around physics' requirements for fusion. There was no attempt to heat the deuterium to millions of degrees or to compress it to high densities. The chemists merely took a little rod of palladium metal, plopped it in a jar full of deuterium-enriched water, and ran an electric current through it. Somehow, without the benefit of high temperature and high pressure, the deuterium atoms were fusing inside that metal.

Though cold fusion seemed ridiculous, physicists could not

dismiss the idea out of hand. It was possible, if unlikely, that palladium metal could somehow force the deuterium nuclei into contact. Pons and Fleischmann could possibly have found a new and fortuitous physical effect that nobody had anticipated. It had happened before. In fact, it had happened before to Fleischmann.

In 1989, Fleischmann was a well-respected English chemist. He had been a key player in the field of electrochemistry, the study of chemical reactions that occur because of the influence of electric currents. He had made his name, in part, by discovering a useful physical effect that nobody had predicted—or, at first, believed. In the early 1970s, he used lasers to detect the presence of a minute amount of a chemical on a piece of silver, even though conventional wisdom said that his results were impossible. The chemical should have been all but undetectable by the technique he used. But Fleischmann was correct; he had done the seemingly impossible. He had unwittingly discovered an effect that would be called surface-enhanced Raman scattering, a phenomenon that is now used in a variety of sensitive chemical detectors. Conventional wisdom was wrong and Fleischmann was right.

The scientific community soon rewarded Fleischmann for his discovery. In the mid-1980s, he was made a Fellow of the Royal Society, the highest honor that Britain bestows upon its scientists. By the late 1980s, his reputation made him welcome at scientific institutions around the world. He spent most of his time hopping between laboratories at his home university in Southampton, the Harwell laboratory (of ZETA fame), and a lab at the University of Utah.

Stanley Pons was the chair of the University of Utah's chemistry department, and the two had a long history together. Fleischmann had taken the younger Pons under his wing in the mid-1970s when Pons was at Southampton. Long after Pons moved back to the States the two kept working together. Fleischmann, the elder statesman, and Pons, the eager young experimentalist, made a good team, producing an enormous amount of research. Pons was particularly prolific. By the late 1980s, he was publishing several dozen papers per year. This was a huge output, and it could be argued that the frantic pace led to careless work. Indeed, over the years, Pons and Fleischmann had published some papers that seemed ludicrous—such as one that involved highly unlikely reactions of nonreactive gases—but the two still maintained a good reputation. This is part of the reason that cold fusion got so much attention. Pons and Fleischmann were established scientists; they were not no-name amateurs like Ronald Richter had been. So when they announced their cold-fusion results in 1989, even skeptical physicists took the claim seriously.

The cold-fusion experiment was deceptively simple. At the heart of each "reactor" was a rod or a sheet made of palladium. Palladium is a whitish metal that shares numerous properties with platinum and nickel. Oddly, it is able to soak up enormous volumes of hydrogen—the tiny hydrogen atoms nestle between the atoms of palladium—so researchers had been studying the metal in hopes of coming up with a method for storing hydrogen in fuel cells.

Pons and Fleischmann had long been intrigued by this hydrogen-sponging behavior. They mused about it—talking while driving across Texas, talking while hiking along a canyon—wondering aloud about what was happening to the hydrogen that was crammed inside the palladium. Perhaps the hydrogens were very crowded in the small spaces between the palladium atoms. Perhaps those spaces were so crowded that the hydrogen atoms were bumping into each other with great force. Perhaps, if the hydrogen was replaced with deuterium…It was a crazy idea, but it just might work. If the pressures inside a palladium cage were high enough, they might just induce fusion in a way that a laser cage or a magnet cage could not. Sitting in Pons's kitchen, the two devised an experiment to discover whether palladium fusion was possible. No law of nature said it was impossible to induce fusion inside a metal cage. It was worth a try, anyhow.

At first, they spent their own money, about $100,000 for the first crude experiments. "Stan and I thought this experiment was so stupid we financed it ourselves," Fleischmann said at the press conference. But by scientific standards, their experimental setup was not that expensive, so it was a worthwhile risk to try, even on their own dime. Sometime in 1984, at least according to their own timeline, they took a chunk of palladium and set it in heavy water, water whose two hydrogen atoms are replaced with deuterium. To the water they added a salt containing lithium and deuterium. Then they stuck in a platinum wire and hooked it and the palladium up to a battery. They hoped that over time the current would cause deuterium to seep into the palladium, where the deuteriums would then begin to fuse. But when Pons and Fleischmann started the experiments, nothing happened. Then, one day, they cranked up the juice and left for the evening. Fleischmann told the *Wall Street Journal* what happened next:

> Sometime during the night the palladium cube suddenly heated up to the point where some of it vaporized, blowing the apparatus apart, damaging a laboratory hood and burning the floor. "It was a nice mess," Mr. Pons said. A check of the laboratory the next day with a radiation counter indicated radioactivity levels three times higher than the normal background levels, apparently the result of a sudden spray of neutrons.*

Pons and Fleischmann took this as a clear sign of fusion. Only a fusion reaction, they reasoned, could vaporize a hunk of metal like that. Mere chemistry could not explain the heat, so that meant something else was going on.

Their crazy hunch had paid off. Pons and Fleischmann felt they had made a momentous discovery. As they continued their research, they tried to keep it secret, letting only a few people into their confidence. But by the late 1980s, they were running out of money, so they started applying for outside funding. The first place Pons turned to was the Office of Naval Research; the ONR was already funding him to the tune of $300,000 per year for other work. But the ONR passed. Next up was the Department of Energy. Its Division of Advanced

---

* As with many of the stories surrounding cold fusion, there are a number of different versions of what happened. For example, some sources say the experimenters (or more precisely, Pons's son) *reduced* the current rather than increased it. The nasty battle that erupted around the research turned the tale of cold fusion into a scientific version of *Rashomon*.

Energy Projects—a group that gives seed money to highly speculative research—was interested. But to award Pons and Fleischmann a grant, the department had to get the application peer reviewed; it had to send the chemists' proposal to other scientists to get their opinions. (Scientific grant proposals, like scientific papers, tend to get accepted only after peer review.) One of the peers who reviewed Pons and Fleischmann's proposal was a physicist at Brigham Young University, Steven Jones. As soon as Jones got a copy, all hell broke loose.

———

Jones was a natural choice for a reviewer. He had long been interested in fusion, particularly fusion under unusual conditions. In the late 1970s, while working at a Department of Energy facility in Idaho, he became intrigued by a bizarre phenomenon discovered by Luis Alvarez—the Oppenheimer critic—in 1956. Alvarez was using a device known as a bubble chamber to study particle interactions; when a particle zipped through the chamber, it would leave a trail of bubbles behind, which allowed physicists to see how particles behaved. He discovered some curious tracks—they had gaps in them—that didn't seem to make sense. He and his group visited Edward Teller's home to talk about the phenomenon, and after an "interesting discussion" Alvarez and Teller realized that the mysterious tracks were the sign of nuclear fusion between a hydrogen and a deuterium.

How could this be? Alvarez's bubble chamber was extremely chilly—not far from absolute zero, in fact. How could the cold, slow-moving deuterium and hydrogen possibly have enough energy to overcome their mutual repulsion and fuse? The secret was a subatomic particle known as the muon. The muon is almost exactly like the electron, but it is some two hundred times heavier than its sibling. Like the electron, it carries a negative charge. And like the electron, it can be captured by a proton to make a hydrogen atom. But this weird muon-hydrogen object is considerably different from ordinary hydrogen. It is more massive—and it is tinier. The muon's extra mass means that it is held much closer to the nucleus than an electron is. Because the muon is held so close to the nucleus, these hydrogen atoms are considerably smaller than

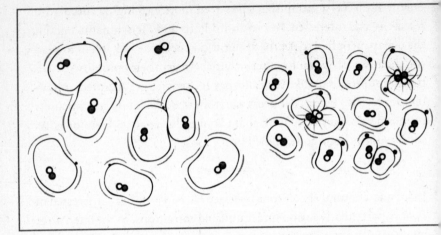

MUON-CATALYZED FUSION: Ordinary atoms have large electron clouds (left) that make it hard for the nuclei to get close enough together to fuse. Replace the electrons with muons (right) and the muon cloud is much smaller; nuclei get together much more easily and are able to fuse at relatively low temperatures.

ordinary hydrogen atoms. Thus when a small muon-hydrogen atom collides with another atom, the two nuclei are much closer together than they would ordinarily be. The muon, wrote Alvarez, "in effect, confines the two nuclei in a small box." Confined in that box, the two nuclei are much more likely to strike each other and fuse.*

Muon-catalyzed fusion, as it came to be known, really *was* room-temperature fusion. If scientists could somehow replace the electrons in a jar full of hydrogen with muons, they would be able to get a fusion reaction without the need for immense heat and pressure; the muon hydrogens would fuse by virtue of their smaller size. Unfortunately, muons are hard to come by. To get them in large numbers, scientists need to build a particle accelerator. Accelerators consume lots of energy, and they are not very efficient.

---

* Apparently, the prolific Andrei Sakharov had predicted this phenomenon in the 1940s.

Even if scientists found an efficient way of producing muons, the muons they would create would last only a few microseconds before decaying into electrons and a handful of other particles. If in those moments the scientists then successfully shot one of those muons into a cloud of hydrogen, they might get lucky and induce two atoms to fuse into helium, but what then? The muon can get trapped in the helium atom, and then it is useless. It will quickly decay without helping any other atoms to fuse. If every available muon catalyzed only one atomic fusion, then there is no hope of producing energy; merely creating the muons and delivering them would consume more energy than was released by that single fusion. If, on the other hand, a muon can escape the clutches of the helium nucleus, then helps another fusion to occur, escapes, helps another fusion, and so forth, then muon-catalyzed fusion would not be hopeless after all. If every muon induces a few hundred fusions before decaying, then perhaps it would be possible to generate more energy than the amount used to create the particles in the first place. Muon-catalyzed fusion would achieve breakeven.

When Alvarez first saw the phenomenon in deuterium, he was extremely excited. "We had a short but exhilarating experience when we thought we had solved all of the fuel problems of mankind," he said in his Nobel Lecture a decade later. "While everyone else had been trying to solve the problem by heating hydrogen plasmas to millions of degrees, we had apparently stumbled on the solution, involving very low temperatures instead." Unfortunately, as Alvarez's team performed more detailed calculations, they concluded that the muons quickly got stuck in helium and decayed, and that muon-catalyzed fusion of deuterium would never lead to a practical energy source. Deuterium-tritium mixtures might fare a bit better, but the outlook was pretty grim.

Jones was more sanguine about the possibility of using muons to generate energy than Alvarez had been, and he sought to prove that muon-catalyzed fusion could, indeed, solve the world's energy problems. With grants from the Department of Energy—the same Division of Advanced Energy Projects that Pons and Fleischmann were soon to get involved with—Jones used an accelerator at Los Alamos to zap deuterium-tritium mixtures with muons. Theory predicted that as the deuterium-tritium mixture got denser, the muon would interact with more atoms before decaying—but that this effect would be very slight. Much to his surprise, when Jones increased the density of his deuterium-tritium mixtures, he discovered that the number of interactions skyrocketed into the hundreds. By 1986, he was claiming to see 150 fusions per muon and predicted that it would be possible to get even more.

If it was true, muon-catalyzed fusion might really become an energy source. In a paper in the prestigious peer-reviewed journal *Nature*, Jones waxed enthusiastic about muon fusion, especially when using a mixture of deuterium and tritium as fuel: "each muon may catalyze hundreds of d-t fusion reactions, releasing a great deal of fusion energy," he wrote, arguing that "muon-catalyzed fusion is an idea whose time has come—again." (He also noted that muon-catalyzed fusion didn't work at high temperatures as conventional fusion did: "The term 'cold fusion' is therefore quite appropriate for the process," he wrote.)

Jones trumpeted the potential of muon-catalyzed fusion in seminars, lectures, and papers, and cowrote a *Scientific American* article about it in 1987. "It is now conceivable that cold fusion may become an economically viable method of generating energy," the article read, and it even included schematics for a "commercial cold-fusion reactor."

Unfortunately, Jones was wrong. His results were not only inconsistent with theory but also with what other groups were finding. A Swiss team, for example, performing similar experiments, was not seeing the same density effects that Jones was observing. Their muons got stuck in helium atoms fairly rapidly, as expected. Instead of seeing hundreds of fusions per muon, they were seeing tens. Muon-catalyzed fusion would never lead to breakeven at this rate. And as the Department of Energy's money for muon-catalyzed fusion began to run out—the Division of Advanced Energy Projects had already spent more than $2 million—prospects for muon-catalyzed fusion began to dim. An outside review by JASON, a secretive group of scientists who advise the government on all matters scientific, put the last nail in the coffin:

muon-catalyzed fusion wasn't worth pursuing, at least as a path to energy. Muon-catalyzed fusion was dead. But cold fusion wasn't.

Around the time that Jones's 1986 *Nature* paper came out, an astronomer and physicist, E. Paul Palmer, attended one of Jones's muon-catalyzed-fusion seminars. The idea of fusion at low temperatures struck Palmer as the possible answer to a conundrum. Palmer was a rogue physicist; he had apparently come to the conclusion that much of what geophysicists believe about the Earth is "a bunch of baloney," and was hard at work formulating alternative geological theories. The conventional wisdom that the Earth's interior is warmed by the decay of heavy elements like uranium struck him as being wrong. And the fact that there is helium-3 in the Earth's crust seemed to him to be evidence of fusion.*

Most physicists would have dismissed Palmer as a crank, but Jones did not. After all, he himself had seen how fusion can happen at low temperatures; perhaps there was some other substance besides muons that could induce low-temperature fusion. Perhaps metals—nickel? platinum? palladium?—could trap hydrogen atoms and force them to fuse. It was cold fusion of another sort. Jones's initial experiments didn't turn up much, despite some halting attempts to capture gamma rays coming from fusion in metal samples. The concept of cold fusion remained on the back burner until the day that Jones received Pons and Fleischmann's grant application in 1988.

There are many different versions of precisely what happened after Jones read the proposal, but there is little doubt that it sparked a race that grew more frantic as each week passed. Pons and Fleischmann's work had much in common with Jones's. Both were hoping to trap deuterium in a hunk of metal—particularly palladium—and force it to fuse somehow. If money was to be made from cold fusion (and if Pons and Fleischmann were correct, cold fusion would be a moneymaker unlike almost any other invention), only the patent holders would see

---

* Most physicists who have studied the matter know that a great deal of helium-3 was trapped in the Earth when it formed, and that a little bit is produced by cosmic rays and by the decay of tritium, not by subterranean fusion reactions.

huge benefits. Only the people who discovered cold fusion would be able to patent the process. And only the people to go public with their work first would be hailed as the discoverers. All the money, glory, and power that might come from the discovery of cold fusion hinged upon being the first to go public. The first to cross the finish line would be hailed as the savior of mankind, as the discoverer of an eternal spring of unlimited energy. The second would become a mere footnote.

Jones, Pons, and Fleischmann had entered an ever-quickening race to run experiments, prove the existence of cold fusion, write a paper for a peer-reviewed journal, and publish it. By early 1989, the competitors had agreed to submit simultaneous papers to *Nature*, so they could all cross the finish line simultaneously. But in a climate of increasing mistrust and antagonism, Pons and Fleischmann jumped the gun. They submitted their paper to the *Journal of Electroanalytical Chemistry* on March 10, and within two weeks they were in front of the microphones, touting their achievement to the world—despite the improbability of what they had found. "Stan and I often talk of doing impossible experiments," Fleischmann said in the official University of Utah press release about cold fusion. "We each have a good track record of getting them to work."

In truth, Pons and Fleischmann did not have the grounds for such hubris. Though they exuded confidence at the March 23 press conference, they already should have known that their data did not add up. They had several lines of evidence for the claim that they had achieved nuclear fusion in their tiny little beakers—but these lines contradicted one another.

The strongest line of evidence, as far as the chemists were concerned, was heat. When Pons and Fleischmann measured the temperature of their apparatus, their electrochemical "cell," they discovered that the palladium was warming it up ever so slightly. Of course, many things can warm up a cell—the electricity they were running through the cell, for example, was certainly contributing to the warming—but Pons and Fleischmann argued that the energy coming from the palladium was considerably more than what they added in the form of electricity.

According to Pons, an inch-long and quarter-inch-thick palladium wire brought water to a boil within minutes, and for every watt of power the scientists put in, four watts came out. More energy out than in implies a reaction of some sort. Since the reaction kept going and going, reportedly for more than one hundred hours, the amount of energy coming from the cell was too large to be explained by a chemical reaction. It was like Marie Curie's hunk of radium; mere chemical processes couldn't seem to explain the heat coming from the cell. To Pons and Fleischmann, this was a smoking gun of a nuclear reaction: fusion.

This sort of evidence would not convince most physicists. To them, the only way to prove that you have achieved fusion is, naturally enough, to show that you are producing some of the by-products of fusion. With deuterium-deuterium fusion, there are a few unambiguous signals that a reaction has taken place.

When two deuterium nuclei fuse ($d + d$), they stick together for a tiny fraction of a second: two protons and two neutrons in a quivering, energetic bundle. Because the conglomerate is so energetic, it cannot hold together completely. One particle is going to pop off and carry away some excess energy. That means either a proton (p) is going to pop off, leaving behind a tritium particle (t) with one proton and two neutrons,

$$d + d \rightarrow p + t,$$

or a neutron (n) is going to pop off, leaving behind a helium-3 nucleus with two protons and one neutron,

$$d + d \rightarrow n + {}^3He.$$

These two branches of the reaction are roughly equally likely: half of the time that you fuse two deuterium nuclei, you will get a proton and a tritium nucleus; the other half, a neutron and a helium-3 nucleus.

Free-floating protons are relatively common, but free-floating neutrons are rarer, as are tritium and helium-3. So if you think that you've got deuterium-deuterium fusion going on in your laboratory, the best way to convince other people is to demonstrate that you are making tritium, helium-3, and neutrons. The neutrons, arguably, should be the easiest to detect. Neutrons penetrate matter very easily, so any neutrons produced by the reaction would quickly fly out through the walls of the beaker and into the walls surrounding the room. A neutron detector need only be placed next to the reactor vessel and it would certainly pick up some of these particles. There are neutrons from other sources—cosmic rays, for example, often produce them in the lab—but luckily the neutrons from deuterium-deuterium fusion have a specific energy.

The fusion of two deuterium nuclei produces a fixed amount of energy: the energy that the particles get from rolling one step down the fusion hill toward the valley of iron. That energy is carried away by the particles created by the reaction. For the branch of the reaction that creates a neutron and a helium-3, the total energy released—in the units that nuclear physicists like to use—is nearly 3.3 million electron volts (3.3 MeV).* That energy is split between the two particles. Furthermore, the heavier particle gets less energy, while the lighter particle gets more.† In this particular case, the heavier helium-3 gets about 0.82 MeV while the lighter neutron gets 2.45 MeV. Every time. So, if you find neutrons flying about with 2.45 MeV of energy, it is a really good sign that you are seeing deuterium-deuterium fusion.

Before the press conference, Pons, Fleischmann, and Jones had

---

* On an atomic scale, a million electron volts is a lot of energy; it is roughly what you would get if you were to take two electron-size particles and convert their mass entirely into energy. For comparison, the combustion of a molecule of TNT yields about 35 electron volts, meaning you would need to burn roughly 100,000 TNT molecules to get the equivalent energy of a single deuterium fusion.

† This is a consequence of the principle known as conservation of momentum, and it's why a motorcyclist suffers so much more than the driver of a car when the two collide. The lighter object (the motorcycle) gets most of the energy of the collision, while the heavier object (the car) gets less. It is the same with subatomic particles. A helium-3 nucleus weighs almost exactly three times as much as the neutron, so the neutron gets three times the energy that the helium nucleus does.

all been looking for neutrons. Jones's team thought it had found a few coming from their experiments—a small, unimpressive bump in a graph. The bump didn't represent a solid discovery; after months of running the experiment, Jones claimed to see roughly twenty neutrons in the 2.45 MeV range. Unimpressive, yes, but Jones considered them a solid sign of fusion reactions. That these neutrons were there at all "provides strong evidence that room-temperature nuclear fusion is occurring at a low rate" in the experiment, Jones later wrote. Pons and Fleischmann had been looking, too, but they were having even less luck. Fleischmann used his Harwell laboratory connections to get a neutron detector, but when they put it near the cell, it didn't show any neutrons. This was a huge problem, because for every watt of power the cell produced, about a trillion neutrons should have been flying out every second. At the power levels Pons and Fleischmann were seeing, their beaker should have been emitting dangerous and easily detectable levels of radioactivity. But it wasn't.

As the days of fruitless searching turned into weeks and the time of the press conference drew closer, Pons and Fleischmann evidently became increasingly concerned. They sent a cell to Harwell to be analyzed with a much more sensitive machine, but the analysis required some time. In the interim, they invited a person from the University of Utah's radiation safety office to the lab to measure gamma rays coming from the cell. The gamma rays, they hoped, would provide an indirect measure of neutrons: when a 2.45 MeV neutron strikes a hydrogen in the water surrounding the palladium, it will emit a gamma ray, again with a very specific energy: 2.22 MeV. The safety officer set up a gamma-ray detector for a few days and collected data. Apparently, Pons and Fleischmann were thrilled with what the machine found, because shortly after analyzing the data, they submitted their paper to the *Journal of Electroanalytical Chemistry* and Utah began setting up the press conference.

When Pons and Fleischmann announced their discovery to the world on March 23, 1989, Utahans immediately sought to capitalize on the news. The day after the press conference, Governor Norman Bangerter announced that he would call a special session of the legislature to appropriate $5 million for cold-fusion research. The appropriations bill passed overwhelmingly. The money would help establish a National Institute for Cold Fusion at Utah. Soon cold-fusion lobbyists would be marching up Capitol Hill seeking tens of millions of dollars, promising that Japan would steal cold-fusion momentum away from the United States if the nation didn't invest immediately.

The scientific community was of two minds. Some were optimistic. Edward Teller called to congratulate Pons and Fleischmann and started a Livermore task force to look into cold fusion. Others, including the University of Utah's own physics department, which had been kept in the dark by the chemists, were extremely wary of the results. No matter the level of skepticism, every scientist wanted details about the experiments, and there were few to be had.

Pons and Fleischmann had held their press conference before publishing their data and their methods. This was very unusual. Scientists communicate through scientific presentations and papers, not through press releases and press conferences. On the relatively rare occasions that a scientific result is important enough to merit a press event, it is usually held at the same moment that the data are revealed to the scientific community through a paper or in a presentation. With the cold-fusion announcement, the paper was missing. No data were available, and scientists had only the scantest details about how Pons and Fleischmann performed their experiment.

Physicists and chemists around the world were frantic; without any data, they had little way to judge whether Pons and Fleischmann were going to solve the world's energy crisis—or whether they were merely full of it. The suspense would last for months.

In the first few days after the press conference, the news seemed good for the two chemists. The press soon learned about Jones's work, and while Jones was much less bold in claiming to generate energy, he, too, was claiming to see fusion in palladium. It appeared to be an outside confirmation of the Pons and Fleischmann claim. No longer could cold fusion be considered the delusion of a single laboratory. As other labs rushed to replicate the experiments, news began to filter in about other confirmations. By early April, researchers at Texas A&M were seeing excess heat in palladium cells; Georgia Tech was seeing neutrons. The University of Washington was seeing tritium. These reports all seemed to provide solid support for cold fusion.

Privately, though, Pons and Fleischmann were getting bad news. Two days before the press conference, Fleischmann learned that even the hypersensitive neutron detector at Harwell wasn't picking up anything. There was no trace of the trillions and trillions of neutrons that should have been flowing from the palladium. Fleischmann apparently explained the discrepancy away, noting that a number of cells that he and Pons had built didn't work; perhaps Harwell was using a defunct cell. It was not a convincing explanation, but it would have to do. But worse news was to come, news that was harder to dismiss.

Four days after the press conference, Pons and Fleischmann began to reveal details of the experiments to some of their colleagues. Fleischmann visited the Harwell lab and gave a seminar on cold fusion. The room was packed with scientists, including some very esteemed ones who had been working with neutrons and gamma rays for years. When Fleischmann showed his gamma-ray measurements to the Harwell crowd, they were shocked. A typical gamma-ray spectrum is a bumpy graph that shows a series of peaks and troughs at various energies, reflecting natural background sources of gamma radiation (such as the decay of radioactive elements). Gamma rays from deuterium should have occurred at 2.22 MeV, right between a gentle peak caused by the decay of radioactive bismuth at 2.20 MeV and a much larger one caused by the decay of radioactive thallium at 2.61 MeV. Instead, Fleischmann showed a ratty little plot that displayed only a single peak, without any nearby landmarks to confirm what the peak really was. Worse yet, Fleischmann was claiming that he was seeing gamma rays that had 2.5 MeV of energy, not the 2.22 MeV that a fusion neutron should emit when it strikes a tub of water.* The peak was in entirely the wrong place.

---

* In fact, a 2.5 MeV gamma ray is obviously wrong. The gamma ray wouldn't be *more* energetic than the 2.45 MeV particle that produced it.

The director of Harwell turned to Fleischmann and said, simply, "It's wrong." Fleischmann wilted. The next day, physicists at the University of Utah—who had been given a preprint of the upcoming cold-fusion paper—told Pons precisely the same thing.

What was going on? Why was the gamma-ray peak in the wrong place? To all appearances, Fleischmann and Pons dismissed the problem, attributing it to a minor error in calculation. When their paper finally came out in the *Journal of Electroanalytical Chemistry*, the lone peak was sitting in precisely the right spot: 2.22 MeV. Perhaps they told the editors about the "error" and corrected it before it was published. However, Pons and Fleischmann apparently failed to spot one occurrence of the old, incorrect value of 2.5 MeV in the manuscript: in the equation where they describe the interaction between a neutron and a hydrogen atom, they declare that the gamma ray would be at 2.5 MeV, not the 2.22 MeV shown by the spectrum.

The problem of the moving peak wasn't public yet, though it soon would be. In the days after the press conference, scientists, still hungry for details about the Pons and Fleischmann experiments, were taking desperate measures. Physicists apparently hacked into Pons's e-mail account looking for clues. One scientist spooked Utah chemists by loitering outside the Pons-Fleischmann lab. A team of plasma physicists at the Massachusetts Institute of Technology resorted to scouring television footage of the lab instruments for data. They succeeded: a broadcast on Utah's KSL-TV showed the entire gamma-ray spectrum, clearly showing the bismuth and thallium peaks. Using that information, they deduced that Pons and Fleischmann's peak had to be near 2.5 MeV as originally presented during the seminar at Harwell, not at 2.22 MeV, as reported in the journal article. Furthermore, even without the television footage, the MIT researchers showed that the Pons-Fleischmann peak was the wrong shape—too narrow and without a distinctive shoulder—for one produced by neutron-created gamma rays. It was a devastating critique, and when Pons and Fleischmann responded to the MIT criticisms in June, the peak had somehow moved back to 2.5 MeV. By that time, most mainstream physicists had already decided that cold fusion was bunk.

However, in late March and early April, the question was still open. While the physicists were still trying to figure out precisely what Pons and Fleischmann had done, the scientific and political communities were dividing into believers and nonbelievers. The biggest critics of cold fusion were plasma physicists. These were the people who knew a lot about the difficulty of achieving fusion, and who had learned through painful experience how neutrons can fool you. They were also the people who had the most to lose if cold fusion worked. Cold-fusion supporters began to sense a conspiracy to attack the Pons-Fleischmann discovery. "There is big money in hot fusion, and if we turn out to be right, hot fusion, I guess, goes away," said the University of Utah president, Chase Peterson. "That represents entire careers, and orthodontia, and college educations for whole families of people that have lived off that dole." In the eyes of supporters, the critics of cold fusion, largely on the East and West Coasts, threatened with obsolescence, were striking at the discoverers of cold fusion in Utah, in the heartland. The university's vice president for research, James Brophy, supported this view: "The black hats, such as they were, came from the hot fusion community.... There was certainly an organized campaign to discredit cold fusion based on the possibility of losing funding."

On the other side, anti-cold-fusion physicists felt that they were simply trying to investigate a very important scientific claim; after all, the whole scientific method relies on the vigilance of the scientific community. Even skeptical fusion scientists, such as Richard Garwin, who helped turn the Teller-Ulam design into a testable bomb, investigated the Pons and Fleischmann claims with an open mind. "Within the next few weeks, experiments will surely show whether cold fusion is taking place; if so it will teach us much besides humility," he wrote in April 1989, even though he himself "bet against its confirmation." But the lack of details from Pons and Fleischmann was frustrating physicists who were trying to confirm the cold-fusion experiments using data gleaned from television broadcasts and newspaper photographs.

Throughout April, the pro-cold-fusion groups had the momentum. Though MIT researchers had reported that they were unable to replicate the experiments in mid-April, there were the confirmatory results on the other side: Jones, Georgia Tech's neutrons, and Texas A&M's heat. When Pons spoke at a hastily cobbled-together special session at the American Chemical Society meeting on April 12, the mood was enthusiastic. The crowd was extremely sympathetic, if for no other reason than the hope that fellow chemists would succeed where physicists had failed. Physicists had spent years trying to harness the power of fusion, noted the American Chemical Society's president in his introduction to Pons's presentation. "Now it appears that chemists may have come to the rescue," he said, triggering applause and laughter. But Pons's presentation generated serious doubts in the audience. Most troubling was when he fielded questions about his control experiments.

If Pons and Fleischmann were actually seeing fusion in a test tube, they should have been able to show that the effect was not due to a quirk in their apparatus. To do this, they needed to run a control experiment—one that was almost identical to the fusion cell, but subtly different in a way that would prevent fusion from occurring. Only then could they prove that fusion was really responsible for the excess heat and other effects they were seeing. In the Pons and Fleischmann case, the obvious control experiment was to run an identical experiment with ordinary water rather than heavy, deuterium-laden water. If deuterium-deuterium fusion was responsible for the excess heat, getting rid of the deuterium and replacing it with ordinary hydrogen should end the fusion and turn the heating off. They then could be assured that the heat had something to do with the deuterium in the beaker. Doing this was absolutely necessary if Pons and Fleischmann were to prove to other scientists that they were not deluding themselves.

Indeed, this sort of control experiment is what budding scientists are taught to do in freshman science classes, and everybody expected it from such established scientists as Pons and Fleischmann—not to have run one would seem absurd. But when questioned about why Pons had not published any control experiments, his reply was cryptic. "We do not get the total blank experiment that we expected," he said. Was he really implying that fusion occurred in the absence of deuterium? This seemed ridiculous even if you accepted that a miracle occurred inside the palladium cell.

At the very least, the scientific community wanted to see the results of those control experiments, but neither the *Journal of Electroanalytical Chemistry* paper nor the one that Pons and Fleischmann submitted to *Nature* had any sign of such a control. "How is this astounding oversight to be explained to students.... And how should the neglect be explained to the world at large?" asked John Maddox, the editor of *Nature*. This was poor science at best, although it was beginning to look much worse than that.

After *Nature* received the twin manuscripts from Pons and Fleischmann and Jones, the journal sent them out for peer review. The reviewers made their suggestions for changes and additional work, and these were sent back to the authors. Jones complied with the reviewers' requests, but Pons and Fleischmann refused to do so, claiming they were too busy with other "urgent work." Though *Nature* emphasized that this did not make the Pons-Fleischmann paper any less believable than Jones's, it was still a deep blow to the team's credibility. Many physicists were beginning to smell a rat, and the rhetoric grew more heated.

The press ratcheted up the rhetoric, too. The *Wall Street Journal* had been enthusiastic about cold fusion since the very beginning. Its reporter, Jerry Bishop, had written a page-1 story covering the Pons and Fleischmann press conference in Utah, and the *Journal* had become the go-to place for optimistic news about cold-fusion developments. When criticism of Pons and Fleischmann began to bubble through the press, especially the liberal press, the *Journal* struck back. In April, the *New Republic* wrote a piece blasting the scientists for releasing the experimental results "in a way that maximized publicity but defied the conventions that are supposed to ensure the reliability of scientific information." The *Wall Street Journal* replied with a caustic editorial linking criticism of cold fusion with other complaints of East Coast liberals: "The pace of scientific advance is sometimes hard to discern amid the unending wail about trade deficits, food chemicals, the ozone layer, the greenhouse effect, animal rights or political ethics," it declared. "Even within the scientific enterprise, the creative impulse of a Fleischmann and Pons must contend today with what might be called 'Academy mentality.'"

The clash between the pro-cold-fusion and anti-cold-fusion camps was becoming an ugly fight. It was also getting more confused by the minute.

The day after Pons's speech at the American Chemical Society, the researchers at Georgia Tech, who had provided evidence in favor of cold fusion, recanted. They weren't seeing neutrons after all. They had made an embarrassing mistake: their detector had been picking up temperature fluctuations rather than neutrons. The Texas A&M scientists also backed off their claims a bit; the amount of excess heat they were seeing had dropped dramatically. Excess heat was still evidence in favor of cold fusion, but the change undermined confidence in the A&M results. Then there was the bizarre claim that Pons and Fleischmann had found helium in their palladium electrodes. In mid-April, the Utah chemists told the press—again, without a formal paper supporting their claims—that their cells were producing helium. But they were claiming the production of helium-4, not helium-3.

If deuterium-deuterium fusion is happening, it is producing helium-3 at a quick rate. Since the branch of the reaction

$$d + d \longrightarrow n + {}^3He$$

happens roughly half the time, one helium-3 should be produced for every two fusions that occur. Helium-3, not helium-4. However, once in a long while—once in about ten million fusions—an unusual reaction does occur. Two deuterium nuclei stick together, producing a helium-4 that is quivering with energy. Usually, the helium-4 can't hold together; either a neutron or a proton pops off. Rarely, though, the helium-4 sheds the excess energy in another way: it emits an enormously energetic gamma ray (with about 24 MeV), and the helium-4 nucleus survives. So the reaction

$$d + d \longrightarrow {}^4He$$

does exist. It is just very rare.

Pons and Fleischmann, backed by the theoretical calculations of two other Utah chemists, were suggesting that this third, rare branch of the deuterium-deuterium fusion reaction had somehow become dominant, suppressing not only the branch that produced tritium but also the branch that produced neutrons. It would explain why the tritium and neutron observations reported thus far were so iffy, and why nobody was spotting helium-3. But physicists weren't buying it. Not only would suppressing the ordinary mechanisms of deuterium-deuterium fusion in favor of this rare branch require a miracle of sorts (nothing like this had been theorized, much less been seen before), but also scientists could point to numerous cases of researchers' being fooled by helium-4 in the atmosphere. In fact, there was a notorious case from the 1920s when two German researchers, Fritz Paneth and Kurt Peters, convinced themselves that a palladium catalyst was turning hydrogen into helium. Instead, the helium they were detecting was contamination from the atmosphere. The case was so similar to the Pons-Fleischmann episode (down to the type of metal used by the experimenters) that it seemed ridiculous for Pons and Fleischmann to rely heavily on helium-4 production as support for their "discovery."

Yet even as the criticism mounted, the researchers betrayed little doubt about their work. On April 26, Pons and Fleischmann, along with Chase Peterson, Steven Jones, and other cold-fusion backers, testified in front of a congressional committee. At stake was a bid to get the federal government to chip in $25 million to cold-fusion research. Pons said he and Fleischmann were "sure as sure can be" that they had achieved fusion, and Fleischmann said he had confirmation of their results from other groups. Even though an MIT physicist urged caution, dubbing the cold-fusion fiasco as "The Case of the Missing Controls," the warnings seemed to fall on deaf ears. Or no ears. The physicist Robert Park noted that "By the time the hearings got around to the skeptics, only two committee members remained, the television cameras were gone."

The physics community was in an uproar. Pons and Fleischmann were too busy to revise their paper for *Nature*, too busy to respond to requests for clarification and information from skeptics, too busy to attend the upcoming American Physical Society (APS) meeting in

Baltimore, but not too busy to hype their claims to Congress in hopes of grabbing $25 million of federal pork. The researchers were making ever more bizarre claims (such as the helium-4 detection) and getting increasingly defensive. In the view of most physicists, the pair had been evasive, self-contradictory, and perhaps less than honest. The mood in the physics community was poisonous. At the Baltimore meeting on May 1, it all erupted.

Neither Pons nor Fleischmann showed up, but Jones, who was not earning the same ire as the other two, was there. Jones was less of a pariah because he had revised his paper for *Nature*, had reported on control experiments with water, and was making much more modest claims than Pons and Fleischmann. And of course, he was appearing before a group of his peers, defending his research. Jones kicked off the session on cold fusion and received a "polite but generally sceptical reception," according to a *Nature* reporter in attendance. Pons and Fleischmann were the main targets. First, Steve Koonin, a fusion scientist at the California Institute of Technology, rubbished the claims of cold fusion—and then he attacked the scientists who made them. "We're suffering from the incompetence and delusions of Professors Pons and Fleischmann," he told the applauding audience. Nathan Lewis, a Caltech chemist, then took up where Koonin left off. He accused Pons and Fleischmann of not stirring the liquid in the cells, allowing hot liquid to accumulate in spots and throwing off their heat calculations. "We asked Pons if he stirred," said Lewis. "No answer." In his rapid-fire presentation, Lewis devastated the Pons and Fleischmann claims. If there was any cold fusion at all—an unlikely possibility—it certainly wasn't the dramatic stuff that the Utah chemists were seeing.

It was a mortal blow. To most mainstream scientists, cold fusion was dead. The *New York Times*'s obituary was a piece entitled "Physicists Debunk Claim of a New Kind of Fusion." Even the *Wall Street Journal* admitted that the session had been a "devastating" attack on the Utah team's credibility, but was less willing to give up hope for cold fusion. (Over the next few weeks, the stream of hopeful news—new confirmations and evidence in favor of cold fusion—continued gracing the pages of the *Journal*.) But to most scientists, cold fusion was well and truly dead, even though, as physicist Park noted, the corpse probably would "continue to twitch for a while." (This was, as it turns out, an understatement.) It was dead to most politicians, too. White House chief of staff John Sununu abruptly cancelled a planned meeting with Pons and Fleischmann on May 4.

The outlook for cold fusion got progressively worse as skeptics piled on, and Pons and Fleischmann got more reclusive and more distant. They failed to attend a cold-fusion meeting later in May. They refused to release an analysis of helium in their palladium rods prepared by the rods' supplier. They even seemed to undermine the research going on at the University of Utah.

Michael Salamon, a Utah physicist who had been running a gamma-ray detector in the Pons-Fleischmann lab, was encountering bizarre roadblocks; the only time that the cells were "working" seemed to be when his equipment was off line. When Salamon wrote a manuscript for *Nature* on his results—entirely negative—Pons's lawyer threatened legal action. And, according to Hugo Rossi, the dean of the University of Utah's College of Science, Pons and Fleischmann didn't cooperate much with the National Cold Fusion Institute, which was established with the $5 million given by the Utah legislature. Speaking about a former Pons postdoc, Rossi explained, "I discovered after awhile that he had instructions from Pons to do nothing [but] set up fake experiments. I discovered this with the help of the assistants who were working for him. [One of them] told me, 'You know, those tubes are running, and there are wires running from them, but they're not hooked up to the computer. Data are not being gathered.'" Pons and Fleischmann lurched toward the fringes of science. But even as they faded from sight, their dream did not die entirely.

———

By the end of May, the mainstream scientific community was convinced that cold fusion was a delusion, and its discoverers, Pons and Fleischmann, were considered either colossally incompetent or patently dishonest. (When the story of the moving gamma-ray peak became widely known, the latter became more and more plausible.) The day after the congressional hearing in April, the Department of Energy asked a panel of scientists (including Koonin) to look into the cold-fusion claims. By the time the draft report came out in July, the verdict was no surprise: there was no convincing evidence for cold fusion. The final report, released in November, was a little more conciliatory, expressing sympathy for "modest support" for well-performed studies to tie up some of the loose ends. There were a lot of them.

Even though Pons and Fleischmann's own work had been thoroughly debunked, a handful of experimenters still thought they had seen heat or tritium coming from palladium cells. Texas A&M's John Bockris and Stanford's Robert Huggins, for example, became staunch supporters of cold fusion based on their labs' results. And, of course, there was Jones. The scientific community found flaws in all these studies. Jones's own cells were shown not to be producing neutrons by a team of physicists led by Moshe Gai, a Yale professor. Huggins was criticized at the APS meeting by a fellow Stanford professor, Walter Meyerhof. Bockris's lab was soon surrounded by intimations of academic fraud, which included spiking cells with tritium from a little bottle. Though the researchers were cleared by a Texas A&M panel, doubts lingered about the quality of their work. This was enough to convince most scientists that cold fusion was not worth any expense of time or effort.

Nevertheless, positive reports from increasingly sketchy research kept dribbling in. These persuaded some scientists, as well as a number of mainstream organizations, including the Electric Power Research Institute and the Stanford Research Institute, that there had to be something to cold fusion. (As late as October 1989, Edward Teller apparently was in favor of funding cold-fusion experiments.) Despite the scorn of most scientists, the research continued to receive money, although it was getting harder to find. University of Utah president Chase Peterson tried to keep the Cold Fusion Institute alive with a $500,000 infusion from his university's research fund.* And so the corpse of cold fusion continued to twitch.

---

* Peterson made the error of calling it an anonymous gift rather than admitting where the funds came from, and he was forced to step down in the ensuing scandal.

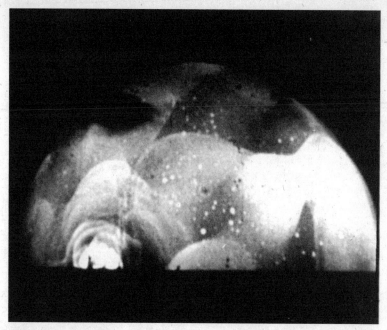

A giant fireball engulfs an island in the initial moments after the Ivy Mike detonation.

The Castle Bravo explosion, much bigger than expected, claimed the first fatality from fusion.

Edward Teller at Livermore

The strikingly round Lake Chagan
was created by the first nuclear blast
of Project No. 7

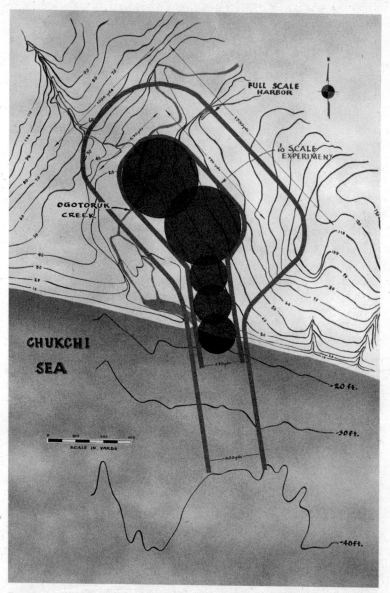

Project Chariot, Teller's plan for carving out an Alaskan harbor with fusion bombs

The Sedan test, a Project Plowshare explosion

The crater left behind by the Sedan test shot was approximately a quarter mile wide and 300 feet deep.

Ronald Richter (LEFT) with Juan Perón, shortly after announcing their surprising fusion project

Stanley Pons (LEFT) and Martin Fleischmann holding one of their cold-fusion cells

Rusi P. Taleyarkhan, one of the codiscoverers of bubble fusion

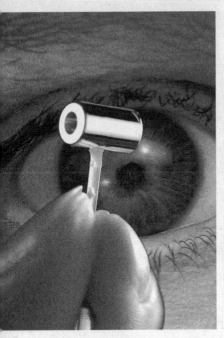
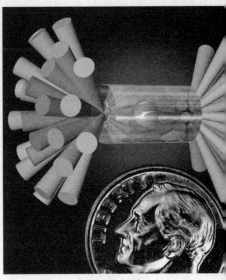

In laser fusion, light strikes a hohlraum, which then emits x-rays that ignite a target pellet inside.

The target chamber at the National Ignition Facility.
During a shot, a hohlraum sits at the very center.

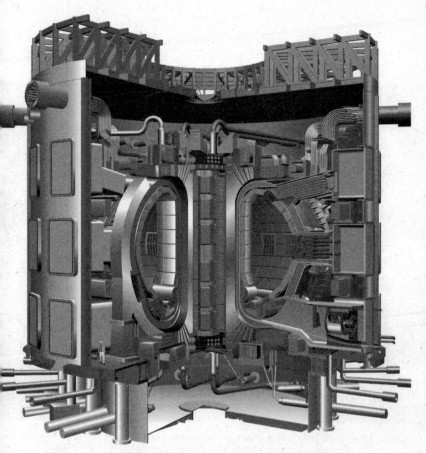

The International Thermonuclear Experimental Reactor will be an enormous—
and expensive—donut-shaped vessel for containing fusing plasma.

Sandia's Z machine in action; it uses electric fields to pinch a target.

Part of what kept the cold-fusion dream alive was the sense of outrage over how Pons and Fleischmann had been treated by the physics community. The smackdown in May had had the air of a public lynching. In its wake, the climate in the physics community had turned from skepticism to scorn. Soon, any cold-fusion believer was ridiculed. It was unseemly, if understandable. A number of people leapt into the fray on the side of the underdogs. The Nobel laureate Julian Schwinger became a cold-fusion supporter and resigned from the American Physical Society in protest over the scientists' poor treatment. Eugene Mallove, MIT's chief science writer, quit his post after alleging that some of the anti-cold-fusion physicists at MIT were engaging in fraud. Mallove then started publishing *Infinite Energy* magazine, which boosted cold-fusion research even as it was getting pushed ever further to the fringes of science.

When Pons and Fleischmann quietly packed their bags and left for France, where a Japanese consortium had set up a cold-fusion research facility, they left behind a small community of true believers. These cold-fusion aficionados supported and encouraged each other, secure in the belief that a revolutionary idea was being crushed by the scientific establishment. There had been a miscarriage of science, they thought. Pons and Fleischmann had been run out of town by the very hot-fusion physicists who were going to lose their funding because of the chemists' discovery.

Pons and Fleischmann continued their research long after the mainstream of science had dismissed cold fusion entirely and had come to consider the whole affair a tremendous embarrassment. The two went their separate ways in the mid-1990s, still insisting they were right, that they had seen excess energy in their palladium cells. The Japanese gave up on cold fusion in 1997, after having spent tens of millions of dollars without any concrete results. The following year, nearly a decade after the scientific community turned its back on the idea, the University of Utah stopped fighting for cold-fusion patents. They were more than $1 million in the hole for lawyer's fees.

Steven Jones, too, was driven to the fringe. Though he kept his post at Brigham Young University, his research got increasingly bizarre. A devout Mormon, he tried to prove that Jesus Christ had visited Mesoamerica (he thought that marks on the hands of Mayan gods were evidence that Christ, with his stigmata, was their inspiration). Then, in 2006, he came out with a study that purported to prove that the World Trade Center had been demolished by explosives inside the building, not by the jets that struck from the outside. BYU initiated a review of the research, and Jones retired from the university shortly thereafter.

If Pons, Fleischmann, and Jones had been the only ones who supported cold fusion, the idea would have long since passed out of the public consciousness. But some serious-sounding scientists at some serious-sounding institutions were convinced that there had to be something to the cold-fusion claims. (Some modern-day cold-fusion work is being done by researchers at the Stanford Research Institute, at a few navy laboratories, and even at MIT.) Some mysterious events also lent credence to the cold-fusion conspiracy theories. In 1992, a researcher was killed in an explosion while performing a cold-fusion experiment, and in 2004, Mallove, the most outspoken proponent of the idea, was found on his driveway, beaten to death.

The cold-fusion movement also drew strength from the press. Reporters seem genetically predisposed to take the side of the underdog, and the cold-fusion-versus-big-science story certainly had one. Some journalists were true believers, and others just were offended by mainstream science's treatment of the cold-fusion researchers. Their gripes came out as a slow and steady drumbeat. "These folks need a fair hearing," said ABC News science correspondent Michael Guillen in 1994. In 1998, *Wired*'s Charles Platt suggested that ignoring new cold-fusion research might be "a colossal conspiracy of denial." The *Wall Street Journal* returned to its cold-fusion roots in 2003 with a column by the esteemed science journalist Sharon Begley: "Cold fusion today is a prime example of pathological science, but not because its adherents are delusional.... The real pathology," she wrote, "is the breakdown of the normal channels of scientific communication, with no scientists outside the tight-knit cold-fusion tribe bothering to scrutinize its claims."

Mainstream physicists saw it differently, of course. Despite the fact that Pons and Fleischmann were claiming something extraordinary—ridiculous, even—the scientific community *had* scrutinized their claims. They found the Utah group's work sloppy at best, and systematically demolished the chemists' claims. Cold-fusion advocates had spent millions of dollars researching the phenomenon and still did not have a device that could reliably heat a cup of water for tea. The burden of proof, as always in science, is on the people who claim extraordinary things. It is their responsibility to perform an experiment so good that it forces the scientific community to abandon its prior beliefs.

This may be the scientific attitude, but it comes across as terribly arrogant, and that served to increase the power of the cold-fusion lobby. By 2004, the pressure had grown to the point that the Department of Energy felt it necessary to review whether cold fusion merited renewed funding. (The term *cold fusion* had been dropped in favor of the less-pejorative *low-energy nuclear reactions*.) The conclusions were much the same as they had been a decade and a half earlier. Yet the mere existence of the review was an indication of the power of the cold-fusion lobby. And the more that people tried to stomp on cold-fusion enthusiasts, the stronger the movement became.

———

The lure of cold fusion and the promise of unlimited, free energy is, itself, a source of power to be reckoned with. Patent examiner Thomas Valone has been under its spell for more than a decade. In 1998, after having worked at the patent office for a few years, he broadcast a plea for cold-fusion aficionados to join him in his line of work, according to *Science* magazine. "Valone called for 'all able-bodied free energy technologists' to 'infiltrate' the Patent Office"—presumably to benefit like-minded cold-fusion and free-energy enthusiasts seeking patents. Valone then attempted to organize what was to be called the First International Conference on Free Energy, which was to be held at the State Department. Mainstream physicists were appalled, including the American Physical Society's Robert Park, who featured the conference in his acerbic weekly *What's New* newsletter. "The speakers list for CoFE is certainly open minded; topics include: assisted nuclear reactions (a.k.a. cold fusion), sonoluminesence (a.k.a. cold fusion), hydrogen technologies (a.k.a. cold fusion), tabletop nuclear transformations (a.k.a. cold fusion)," Park wrote, noting that the conference was to be held under the "auspices of the U.S. State Department in the Dean Acheson Auditorium." An embarrassed State Department booted the conference, but the meeting survived. Within a few weeks it was renamed the First International Conference on Future Energy, and it had apparently found another home. Valone billed the event as being held "In cooperation with the U.S. Department of Commerce"—the department that runs the U.S. Patent and Trademark Office. Commerce kicked the conference to the curb. Valone, too.

In May 1999, a week after Valone's conference took place in a Bethesda hotel, his supervisors started a process to remove him from his job, alleging, among other things, that he had misrepresented the Commerce Department's role in the conference. By the end of August, he was fired—but Valone filed a grievance. He felt he was being unfairly persecuted for his beliefs.

When the case was heard, the arbitrator had harsh words for the patent office and its reliance on hearsay, and failure to follow proper procedure. But the harshest criticism went to the physicists who had attacked Valone. It is easy to understand why Park and his American Physical Society colleagues went after cold fusion, he wrote:

> The federal government's budget pie for research and development in the areas of theoretical physics and chemistry is limited and, by and large, only traditional physicists represented by organizations like the APS, and its counterpart for conventional chemists, have been invited to sup on that pie. The last thing they want is any new guests invited to the table.

The arbitrator's words echo those of the cold-fusion community. Small-minded physicists are trying to suppress research that would take money away from their own endeavors. Valone got his job back along with back pay.

The second Conference on Future Energy, with its lightning-bolt

BioCharger, its talk about antigravity stealth bombers, and its whole-hearted embrace of cold fusion, was Valone's victory celebration. It represented the defeat of the forces trying to suppress his views and the comeuppance for the physicists who had hounded Pons and Fleischmann abroad and driven cold fusion to the fringe. It was 2006, nearly two decades after the two chemists had been ridiculed by mainstream scientists, but the gathering proved cold fusion was still alive. The dream of unlimited fusion energy in a room-temperature test tube was too powerful for mere science to destroy.

CHAPTER 7

SECRETS

Everything secret degenerates, . . . nothing is safe that
does not show how it can bear discussion and publicity.

—LORD ACTON

The cold-fusion affair captured the imagination of the public. Two chemists, two outsiders, claimed to have succeeded with a cheap, tabletop experiment where legions of physicists with hundreds of millions of dollars had failed.

Fusion energy is hard. Even if you manage to get a fusion reaction going in a small device—and a number of people have succeeded in doing just that*—"tabletop" devices all consume more energy than they produce. The more researchers experiment with fusion, the more most of them are convinced that the best—if not the only—way to create a fusion reactor is with a hot plasma, confined and compressed by some powerful force. Nowadays, that leaves only two realistic options: big, expensive magnets or big, expensive lasers.

Both approaches require billions of dollars and thousands of scientists. And both have secrets. Laser fusion's secret is a matter of national

---

* See appendix.

security; magnetic fusion's secret is a matter of some embarrassment. Both secrets threaten the future of fusion energy.

———

In the late 1970s, before Shiva came on line, laser scientists at Livermore were extremely confident that they were on the fast track to fusion energy. They believed that Shiva, with its twenty beams, would likely achieve breakeven: the machine would produce as much energy in fusion as was poured into the system by its lasers. They were sure that they would produce a fusion reactor by the century's end, and they were not ashamed to tell the press about it. In 1978, shortly before all the Shiva experiment's lasers were all turned on, Livermore's physicists were talking to the press about having a fusion power plant working in the late 1980s or early 1990s.

Despite the numerous problems that the physicists were encountering with laser fusion—the loss of energy to electrons, the Rayleigh-Taylor instability, and numerous other effects that made it harder than expected to compress and heat a deuterium pellet—they felt that they had good reason for optimism. It was called LASNEX.

LASNEX is a very intricate computer program meant to simulate what happens in the heart of a laser fusion experiment, and though the computer program is still classified, a few details are available outside the fusion community. Scientists apparently began working on it in the late 1960s or early 1970s. When Livermore's John Nuckolls wrote about laser fusion in *Nature* in 1972, he referred to an early version of the code, and even then it was relatively advanced. LASNEX described every possible interaction of light, electrons, and nuclei that the designers could imagine. It told them how a cold pellet of matter begins to compress under laser light or x-ray emissions, how hot electrons bleed off energy, how fusion in the belly of the pellet causes it to expand. Physicists could tinker with various conditions, changing the size or contents of the pellet, the color and intensity of the light that shines upon it, the number and location of laser spots hitting a target or a hohlraum. They could, in short, run lots and lots of laser experiments before ever building a real-life laser device. To laser fusion scientists, LASNEX was a secret weapon. Nuckolls likened LASNEX to "the breaking of an enemy code. It tells you how many divisions to bring to bear on a problem."

In the late 1970s, the LASNEX simulations told Nuckolls and the Livermore team that not that many divisions—not all that much laser power—needed to be brought to bear on the pellet to reach breakeven. The Shiva laser system, with its twenty beams, could pour about 10,000 joules into a pellet within a billionth of a second.* According to LAS-NEX, this energy should be enough to ignite the pellet, and the reaction would produce as much energy as it consumed.

LASNEX was wrong. Very wrong. After a year's worth of experimentation, the *New York Times* reported that "energy release at the giant $25 million Shiva fusion machine at the Lawrence Livermore National Laboratory in California fell short of the more optimistic estimates by a factor of 10,000." LASNEX was a quite a few divisions short of a victorious army.

What went wrong? Because LASNEX is classified, it is hard to tell for certain. But what is certain is that a computer simulation is only as good as what its programmers know. When they write the code, they try to include everything they can think of, but there might be, and usually are, unanticipated phenomena lurking around the corner. It's probable that as researchers experimented with ever more powerful lasers, they discovered difficulties that they didn't expect and that weren't programmed into the code. LASNEX wasn't a perfect reflection of reality, because its programmers didn't have perfect understanding.

Furthermore, even if every possible phenomenon were somehow programmed into LASNEX, the code wouldn't necessarily predict the

---

* A joule is a measure of energy that is related to the watt, which is a measure of power. A 100-watt lightbulb, for example, consumes 100 joules of energy every second. Ten thousand joules isn't a lot of energy—the amount consumed by a 100-watt lightbulb in less than two minutes—but poured into such a tiny space and over such a short time it becomes a considerable amount of power. And it takes much more than 10,000 joules of energy to produce a laser beam that powerful.

actual progression of a given experiment. There is a built-in limitation in the LASNEX code: it is only two-dimensional.

Running a LASNEX simulation takes an enormous amount of computing power; even the simplest simulated experiments require supercomputers to chug away for a long time, moving imaginary electrons and nuclei and photons around in vast memory banks. The calculations are so complex, in fact, that a full-scale, three-dimensional simulation was way too much for the computers to handle. Instead, the programmers decided to simulate only a flat, two-dimensional slice of an imploding pellet. It was a necessary simplification; it brought the calculations into the realm of possibility. But at the same time, it meant that the LASNEX code couldn't do a true simulation. If the real three-dimensional implosion behaved even slightly differently from a two-dimensional mockup, then LASNEX could not predict its behavior very well.

Nevertheless, LASNEX's programmers were justly proud of their accomplishment. Under many circumstances, the code did predict precisely the outcome of a given experiment. However, once Shiva became fully operational, it became dreadfully apparent that LASNEX didn't give quite the right answer all the time—and its promises of success didn't come true. Its failed predictions were a blow to the Livermore scientists, but they refused to be derailed. A much bigger project, Nova, was already under way. The LASNEX code—presumably tweaked to take into account the results from the Shiva experiments—apparently predicted a resounding success with Nova, which was ten times more powerful than its predecessor.

Then Nova, too, failed to achieve breakeven. The laser was certainly generating fusion reactions. By the mid-1980s it was achieving about ten trillion fusion neutrons with each shot. But again, the laser consumed one thousand to ten thousand times as much energy as the fusion reactions produced. Once more, LASNEX had failed, and the scientists' optimistic expectations were crushed. This time, though, their failure had cost almost $200 million.

Even though LASNEX has failed and failed again, the Livermore scientists still insist that they have good reason to trust the computer's more recent predictions. They claim to have experiments that back up the computer code, but it is impossible to tell, from the outside, whether they are telling the truth. The evidence they cite, like the LASNEX code itself, is classified.

Beginning in the late 1970s and extending until the late 1980s, the national laboratories at Los Alamos and Livermore embarked on a classified program of experiments dubbed Halite/Centurion. These experiments were intended to help the laser fusion community test LASNEX and improve the code, and to determine, once and for all, the conditions required to ignite a fusion reaction in a pellet of deuterium. Instead of using lasers to ignite a pellet, the Halite/Centurion program used nuclear bombs.

Though very little information is available about the Halite/Centurion experiments, some details have dribbled out. It appears that the tests used hohlraums—the little metal tubes that are crucial to indirect-drive laser fusion—containing target pellets. These hohlraums and pellets were placed deep underground, at various distances from nuclear bombs. When the bombs went off, they radiated x-rays in all directions. Some of those x-rays shined into the hohlraums, which reradiated x-rays toward the pellet, just as in a laser fusion experiment. These reradiated x-rays, in turn, crushed the deuterium pellets, and scientists observed the resultant fusion reactions.

Laser fusion scientists state that the Halite/Centurion tests were a ringing confirmation of their beliefs and of LASNEX's predictions. They claim that the tests put to rest questions about the feasibility of laser fusion reactors—though they don't give any details. If the pro-laser-fusion scientists are to be believed, then Halite/Centurion showed that the laser fusion program is on the right track.

Not everybody agrees. Apparently, the hohlraums in the Halite/Centurion experiments received varying amounts of energy, from tens to hundreds of millions of joules, about a thousand times greater than the energy even the Nova laser would deliver. But even with that much energy driving them, 80 percent of the capsules failed to ignite, says Leo Mascheroni, a former Los Alamos laser physicist. Worse yet, he says, LASNEX didn't predict the failures. Mascheroni argues that the pro-laser-fusion lobby is hiding negative results behind a wall of secrecy; if outside scientists could see the data, he says, they would conclude that Halite/Centurion proved that the laser fusion program was failing miserably.

Who is correct? It's a secret. Those scientists who have access to the data from Halite/Centurion can't talk; it's unlawful for them to make any details public. Those who don't have access obviously can't assess the arguments. It's the big secret of laser fusion. Only the scientists working on laser fusion can see the proof that they are on the right track. Those of us on the outside are forced to take their word for it. And for the past few decades, their word hasn't been very good at all.

———

Magnetic fusion has the advantage of openness. You can read almost all the literature that has been written about it. You can visit the facilities and walk around without fear of stumbling into a classified area. Indeed, by the 1990s some fusion labs looked as if they were desperate for visitors.

It was a far cry from the golden age of fusion. Twenty years earlier, in the mid-1970s, fusion had plenty of support from Congress and from the public.* The OPEC crisis had sent fusion budgets soaring, and scientists planned large magnetic fusion machines around the country. Most of them were tokamaks, but a few other designs were also planned, such as a mirror-type machine at Livermore.

The big tokamak in the United States would be at Princeton: the Tokamak Fusion Test Reactor (TFTR), which promised to achieve breakeven. TFTR was supposed to cost a bit more than $300 million,

---

* Some of the support was unwanted. The conspiracy theorist and presidential candidate Lyndon LaRouche was a big supporter of fusion power, and founded an organization, the Fusion Energy Foundation, to support the cause. According to LaRouche's *Executive Intelligence Review*, the foundation was shut down in the late 1980s after a long campaign of harassment by British and Israeli intelligence operatives. According to most other sources, the foundation was shut down when the government went after LaRouche for mail fraud. (He was convicted in 1988.)

but as often is the case with cutting-edge science projects, the expenditures ballooned well beyond that by the time the project was finished. But achieving breakeven was the minimum requirement for a fusion reactor, and that made the newest generation of tokamaks big and expensive. The United States was not the only country willing to spend hundreds of millions on big tokamaks: European countries were banding together to build one known as the Joint European Torus (JET), and Japan was planning to build a tokamak that would be known as JT-60. The three devices were similarly enormous and expensive. (They had some design differences, too. TFTR would be able to produce higher magnetic fields, and JET would be able to induce larger currents in the plasma; the JT-60 fell between the two extremes.)

In the late 1970s, morale in the magnetic fusion community was extremely high. Though they were still far away from breakeven—fusion energies were still ten thousand times smaller than the energy put in—they had been making steady progress over the years. As the machines got bigger and more expensive, scientists were able to get higher temperatures and densities in their plasmas, and to hold them for longer times. Physicists were confident that the new, large tokamaks being built would achieve breakeven, and perhaps go beyond. So were politicians. In 1980, President Jimmy Carter signed into law an act that promised to double the fusion budget in seven years—from nearly $400 million annually—and established the national goal of "the operation of a magnetic fusion demonstration plant at the turn of the twenty-first century." The promised land was in sight. It would take only twenty years to get there.

For a tokamak, the promised land is not just breakeven. It is known as "ignition and sustained burn." Unlike laser fusion devices, which have to create individual bursts of fusion energy, a magnetic fusion device like a tokamak can, in theory, run nonstop, producing continuous energy. Once scientists are able to get their magnetic bottles strong enough, they will be able to exploit this and keep a fusion reaction running indefinitely. The fusion reactions in the belly of the tokamak should suffice to keep the plasma hot, so after they get it started, the reaction will essentially run itself. All the scientists have to do is periodically inject some more deuterium and tritium fuel into the reactor and remove the helium "ash" from the plasma. Once you figure that out, you've got an unlimited source of power. Ignition and sustained burn are much better than mere breakeven: once you've got it, you've built a working reactor. And Carter's plan called for developing just that.

By the time Ronald Reagan came into office, the climate for fusion was already changing. The OPEC crisis was fading into memory, and energy research was not a high priority for the new president. He scuttled Carter's plan, and as budget deficits rose, fusion energy money began to disappear, $50 million hunks at a time. The panoply of glorious experiments planned in the 1970s began to crumble under increasing financial pressure. As magnetic fusion budgets dwindled, researchers struggled to save their precious tokamaks from the budget ax. A huge magnetic-mirror project that had already swallowed more than $300 million was scrapped just as it finished its eight-year construction and was about to be dedicated.* It never got turned on. One after another, new facilities—such as the "Elmo Bumpy Torus" and the "Impurity Studies Experiment"—died on the drawing board. The TFTR program was delayed, but not cancelled. The big tokamak was barely able to keep itself alive; everything else starved. With budgets in free fall, there was no room for anything other than the tokamak program, and even that was in jeopardy.

Despite the budget crunch, TFTR (and JET, JT-60, and a handful of other tokamaks worldwide) was steadily closing in on breakeven. It was holding plasmas for seconds at a time and achieving temperatures close to a hundred million degrees. Even with the improvements, breakeven was still a distance away, and the promised land of ignition and sustained burn was clearly out of reach. There was no way, with budgets as they were, that fusion scientists could ever hope to build a magnetic fusion reactor. A tokamak big enough and powerful enough to keep a plasma burning indefinitely would cost billions, and America's fusion budget could never withstand that sort of strain. The story was little different overseas. No single nation could afford to build a tokamak that could

---

* The dedication ceremony went on as planned, even though the project had been cancelled weeks before. "I thought I was going to a wake," said one of the ceremony's attendees.

achieve breakeven and sustained burn. Perhaps, though, by pooling their resources and joining together in one great effort, fusion scientists around the world could finally build a working fusion reactor.

The idea of an international reactor had been around since the budgets started dropping, but it truly came to life in 1985. At a summit in Geneva, Reagan and the Soviet leader Mikhail Gorbachev tried to reduce tensions between the U.S. and the USSR. Gorbachev suggested to Reagan the possibility of a joint effort to build a fusion reactor. Reagan jumped at the chance, as did France and Japan. Together, the four countries would build an enormous tokamak that would finally achieve ignition and sustained burn. For the first time, humans would be able to harness the power of the sun for peaceful purposes. The International Thermonuclear Experimental Reactor (ITER) was born.

ITER was to be a monster. As design work began on it, scientists realized that it would cost $10 billion. The four parties, working together, could cough up the money, but ITER would devour the fusion budgets of all the participating countries.* Even the big tokamaks—TFTR, JET, JT-60—would not survive. Once the ITER project was under way, there would be no room in the budget for anything else. This was a big problem.

Princeton scientists did not want their facility to disappear. Other fusion researchers, especially those who thought that non-tokamak machines were still worth exploring, were angry that the world was going to gamble all its fusion money on a tokamak while ignoring all other possibilities. Almost everyone agreed that a big international reactor effort would be a wonderful thing, but at the same time everyone wanted to have a thriving domestic fusion program, too. Fusion researchers wouldn't get both, especially with the budgets dropping precipitously. In the early 1990s, with ITER in ascendance, the Princeton Plasma Physics Laboratory seemed marked for death.

The first thing that would strike a visitor to the Princeton facility in the early 1990s would be the circles. There were circles everywhere. In

---

* Some wags noted that ITER (pronounced "eater") was a frighteningly apt name for the device.

the lobby, an office assistant swiveled about behind a large ring-shaped desk. A circular sofa surrounded a donut-shaped model of the TFTR. Other models of ringlike tokamaks were displayed in the waiting room. Even the auditorium was semicircular. And of course, the heart of the whole facility was the donut-shaped TFTR tokamak.

The second thing that would strike a visitor was the air of quiet desperation that hung about the lab. The staff was trying to sell fusion to the public, and while the TFTR was setting temperature records almost daily, nobody seemed to be buying. Budgets were still dropping, and the taxpayers didn't protest. The lab, quietly, tried to change that attitude. Along each wall of the laboratory's lobby, colorful posters exhorted the taxpayer to back fusion research. "Why Fusion?" read one. "Do We Really Need To Spend This Much On Energy Research?" asked another. Rush Holt, a physicist and the spokesman for the TFTR project, promised great things for TFTR—6 watts out for every 10 put in, within spitting distance of breakeven—but most of all, he conjured a future with fusion energy. Without it, he said, humanity would be in trouble.*

Where can we as a society get our energy? Fossil fuels pollute, cause global warming, and are running out. Renewable sources—solar, geothermal, wind—can't provide nearly enough energy for an industrial society.† That leaves nuclear energy: fusion or fission. Holt argued that fission is messy: a fission reactor uses up its fuel rods and leaves behind a radioactive mess that nobody knows how to dispose of. Fusion, on the other hand, leaves no harmful by-products. It runs on deuterium and tritium, he said, and leaves only harmless helium behind. Clean fusion energy would be a much better choice.

---

* In 1999, Holt was elected to Congress as the representative for Princeton's congressional district.

† At one point, Holt displayed a graph showing life expectancy and energy consumption: the more energy a society consumes, the longer its people live—implying that societies should boost their energy output. Of course, this is a causation-correlation fallacy. A highly industrialized society consumes more power, and it has longer life expectancies because it has better medical care. The increased power consumption doesn't *cause* the increased life expectancy, even though the two are *correlated*.

This is the sales pitch of faithful magnetic fusion scientists everywhere. Fusion provides unlimited power—clean, safe energy without the harmful by-products of fission. But there is a dirty little secret. Fusion is not clean. Once again, it's the fault of those darn neutrons.

Magnetic fields can contain charged particles, but they are invisible to neutral ones. Neutrons, remember, carry no charge and do not feel magnetic forces. They zoom right through a magnetic bottle and slam into the walls of the container beyond. Since a deuterium-deuterium fusion reaction produces lots of high-energy neutrons (one for every two fusions), the walls of a tokamak reactor are bombarded with zillions of the particles every moment it runs.*

Neutrons are nasty little critters. They are hard to stop: they whiz through ordinary matter rather easily. When they do stop—when they strike an atom in a hunk of matter—they do damage. They knock atoms about. They introduce impurities. A metal irradiated by neutrons becomes brittle and weak. That means the metal walls of the tokamak become susceptible to fracture before too long. Every few years, the entire reactor vessel, the entire metal donut surrounding the plasma, has to be replaced.

Unfortunately, neutrons also make materials radioactive. The neutrons hit the nuclei in a metal and sometimes stick, making the nucleus unstable. The longer a substance is exposed to neutrons, the "hotter" it gets with radioactivity. By the time a tokamak's walls need to be replaced, they are quite hot indeed.

Though fusion scientists portray fusion energy as cleaner than fission, a fusion power plant would produce a larger volume of radioactive waste than a standard nuclear power plant. It would also be just as dangerous—at first. Much of the waste from a fusion reactor tends to "cool down" more quickly than the waste from a fission reactor, taking

---

* If the fuel is a mixture of deuterium and tritium, the results are even worse; there are more neutrons, and they have more than five times as much energy: 14.1 MeV instead of 2.45 MeV. Advanced fuels, such as ones based on fusing boron-11 with protons or helium-3 with itself, could reduce this neutron problem dramatically, but right now, these are a pipe dream.

a mere hundred years or so until humans can approach it safely. But it means that humans will have to figure out where to store it in the meantime, as well as the rest of the waste that, like spent fission fuel, will remain untouchable for thousands of years. Fusion is a bit cleaner than fission, but it still presents a major waste problem.

Fusion scientists recognize this, of course. They are working on exotic alloys that are less affected by neutron bombardment, materials made of vanadium and silicon carbide. However, developing those materials is going to cost a lot of money, and they will still present a waste problem, albeit a reduced one.

It's an open secret. Fusion isn't clean, and it probably never will be.

CHAPTER 8

BUBBLE TROUBLE

Hegel observes somewhere that all great incidents and individuals of world history occur, as it were, twice. He forgot to add: the first time as tragedy, the second as farce.

—KARL MARX, *THE 18TH BRUMAIRE OF LOUIS NAPOLEON*

The mere mention of cold fusion made everyone bristle. The scientists, the press office, the editor of the magazine all objected to anyone's using the term. But the phrase was soon echoing across the nation. It was on the front pages, in the evening television broadcasts, and plastered all over the press. Cold fusion rides again. History seemed to be repeating itself.

The controversy seemed familiar from the start. Scientists at Oak Ridge National Laboratory and Rensselaer Polytechnic Institute, both well-respected institutions, claimed that they had created fusion in a little beaker of acetone not much bigger than the original Pons and Fleischmann cell. In many ways, though, this situation was very different from cold fusion. Rather than announcing their results at a press conference, the scientists sent them to *Science* magazine, the most prestigious peer-reviewed journal in the United States, and their paper had been accepted. The scientists weren't saying they had discovered dramatically new physics, as Pons and Fleischmann's palladium-catalyzed fusion would have required. These bubble fusion reactions were supposedly happening at tens of millions of degrees, rather than at room temperature.

But like cold fusion, the bubble fusion researchers believed their work could lead to an unlimited source of energy. And like Pons and Fleischmann, the bubble fusion scientists quickly came under attack by some of the leading fusion physicists in the nation. Even before their paper had been published in *Science*, the bubble fusion scientists were labeled as incompetent. It got worse after publication. Increasingly isolated, they were forced toward the fringe, and before long they were fighting accusations of scientific misconduct and a fraud investigation that led all the way to Capitol Hill. History *had* repeated itself.

The bubble fusion imbroglio was a twisted reflection of the cold-fusion affair. The second time around, the tale would be a tragedy as well as a farce. Researchers, peer reviewers, editors, journalists, press officers, and all the other players in the drama were caught in a colossal web of mutual misunderstanding. It was a story of good intentions gone wrong, of paranoia and mistrust, and of hubris that led to the downfall of a scientist.

When the bubble fusion story broke in 2002, I was a reporter for *Science*, ground zero for the controversy.

Science is famous because it is arguably the premier peer-reviewed scientific journal in the United States. For many scientists, a publication in it (or its British rival, *Nature*) would be considered a major coup, perhaps even the crowning achievement in an average scientific career. Researchers from around the world submit manuscripts to *Science*, and it is an enormous task to examine the submissions and select those worth publishing.

I had nothing to do with the peer-reviewed section of *Science*. I worked for the news pages at the front of the magazine. News reporters at *Science* are deliberately isolated from the peer-reviewed section. We weren't told about manuscripts in the pipeline, or about the status of a paper undergoing peer review. We weren't even allowed to know who the peer reviewers of a given manuscript were.* So I was quite surprised when, on February 5, 2002, my editor, Robert Coontz, e-mailed me a paper entitled "Nuclear Emissions during Acoustic Cavitation." It had already been through the peer-review process, but it wasn't an ordinary manuscript. "Here's a *Science* paper that's likely to be **very** controversial," Coontz wrote. "First task is to decide whether we want to cover it." Within a few seconds, I knew it was going to be explosive.

The manuscript was couched in the typical cold, technical language of the scientific paper, but its authors, a team led by Rusi Taleyarkhan at the Oak Ridge National Laboratory, were making a claim that seemed eerily reminiscent of cold fusion. They claimed to have induced fusion reactions on a tabletop using a process that might lead to energy production. More important, they did it in an ingenious, and seemingly plausible, way. They did it with a technique linked to a mysterious phenomenon known as sonoluminescence.

As early as the 1930s, scientists had discovered a bizarre method to convert sound into light. If you take a tub of liquid and bombard it with sound waves in the correct manner, the tub begins to generate tiny little bubbles that glow with a faint blue light. This phenomenon is not perfectly understood, but scientists are pretty sure they know what is going on, at least in gross terms.

If you have ever belly flopped off a diving board, you know that a liquid like water doesn't always behave quite like a fluid. Hit it hard enough and fast enough, faster than the water can flow out of your way, and it feels almost like concrete. It behaves more like a solid than like a liquid. This is more than a mere metaphor. Under certain circumstances—if you hit a liquid in the right way—it will "crack" just as a solid would. The liquid ruptures, creating tiny vacuum-filled bubbles that

---

* That way, when we reported about a manuscript that was appearing in *Science*, as we often did, we would not be influenced by the opinions of the editorial side of the journal. We could criticize an article in our own magazine's pages or refuse to cover it, regardless of the opinions of *Science*'s scientific editors.

instantly fill with a tiny bit of evaporated liquid. This phenomenon is known as cavitation, and it occurs in a number of different places. Submarine propellers, for example, cause cavitation if they spin too fast. Sound waves rattling through a fluid can also create these bubbles.

Under the right conditions, the sound waves reverberating through the liquid also cause these bubbles to compress and expand, compress and expand. Each time the bubbles are squashed by the sound waves, they heat up. If the sound waves are just right, the bubble can collapse to roughly one-tenth its original size, heating up to tens of thousands of degrees and emitting a flash of light. This is sonoluminescence.

Taleyarkhan wondered what was happening at the center of those collapsing bubbles. What would happen if you replaced water with a deuterium-laden liquid? If those bubbles got squashed far enough and became hot enough, could they induce the little bit of deuterium vapor in the center of the bubble to fuse? Could they induce a fusion reaction in a beaker?

The first problem he encountered was that tens of thousands of degrees isn't nearly enough to induce fusion, so ordinary sonoluminescence didn't have any hope of getting deuterium nuclei to stick together. For fusion, Taleyarkhan needed to heat deuterium and to tens of *millions* of degrees, a thousand times hotter than what traditional sonoluminescence could achieve. The only way to get those temperatures was to compress the bubbles far more than had ever been done before, either by squashing them tighter or by starting with bigger bubbles. Taleyarkhan had figured out an innovative way to do the latter.

His research team started with a solution of deuterated acetone, the same molecule that's in nail polish remover, except for the fact that its six hydrogen atoms have been replaced with deuteriums. Then they irradiated the liquid with energetic neutrons and exposed it to sound waves. The energetic neutrons poured their energy into the solution and birthed very large bubbles—tens or hundreds of times larger than the ordinary bubbles in sonoluminescence—and, according to Taleyarkhan and his colleagues, the sound waves compressed them by a factor of ten thousand. This was a much higher compression than had ever been observed before. Taleyarkhan's calculations implied that this extreme compression led to a temperature in the range of millions of degrees. This, in turn, supposedly led to fusion.

To all appearances, Taleyarkhan and his colleagues did all the right things when they went looking for deuterium-deuterium fusion. The paper told of how the researchers looked for neutrons—and found them. Tritium? Found it. They also avoided many of Pons and Fleischmann's mistakes. They ran the obvious control experiments, substituting ordinary acetone for the deuterated variety. The neutrons and tritium disappeared. Finally, the paper convinced a science editor and a group of peer reviewers who, presumably, were satisfied with its quality.

But I was skeptical. For one thing, I knew Taleyarkhan, and while I held him in reasonably high esteem, I didn't think of him as a fusion expert. A few years earlier—in 1999, when I was a reporter for *New Scientist* magazine—I had written about one of his inventions. He had figured out a clever way to make a gun that would shoot bullets at different speeds. In theory, you would be able to turn a dial on a gun and set it to "stun" with low-velocity bullets or to "kill" with high-velocity ones. (It used an aluminum-based propellant that could do things ordinary gunpowder couldn't.) Interesting stuff, but not the sort of thing a fusion expert would invent. Taleyarkhan was a nuclear engineer, and I associated him with steam explosions and propellants and reactor safety, not fusion physics. What really bothered me, though, were the neutrons.

The bubble fusion paper was going to live or die by the neutrons Taleyarkhan was claiming to see. Neutrons were what killed Pons and Fleischmann. Neutrons were what killed ZETA. Without a nice, clear demonstration of neutrons of the proper energy—2.45 MeV—streaming from the experimental cell, nobody would take Taleyarkhan seriously for a minute. So the first thing I looked at was the paper's graph of neutrons. I was surprised.

Skeptical physicists would only be convinced by a detailed graph showing how many neutrons were detected at what sorts of energies. Taleyarkhan's paper had a few graphs, but they were far from detailed. The main one only had four points—two for the deuterium experiment and two for the control experiment—telling how many neutrons were detected above and below 2.5 MeV. That wasn't *nearly* enough, at least in my opinion. I expected a neutron spectrum to have tens of points, not two. Without that level of detail, I didn't think that there was enough information to determine whether the experimenters were seeing something real.

I was uneasy. The content of the graph did not rule out the claim of fusion. Taleyarkhan's team may well have seen neutrons that were drop-dead evidence of fusion. But if they did, I couldn't tell from the graph. If they had confirmatory data, they were not presenting it in a convincing way. If they didn't know how to convince other scientists of their claims, I suspected that they didn't know enough about the field to make such claims in the first place.

That was my initial impression. But as a journalist, I've learned that first impressions are very often wrong. In fact, I wanted to be convinced that I had erred in my snap judgment, if for no other reason than I thought it would make a better story if Taleyarkhan was correct. Furthermore, I knew that the manuscript had gone through *Science*'s peer-review process. The editor who had handled the manuscript—I presumed that it was our physics editor, Ian Osborne—did not laugh it out of the room when he read it. The peer reviewers (whose identities I didn't know) had also, presumably, vetted the manuscript and found it worthy of publication. This certainly did not ensure that Taleyarkhan and his colleagues were right, but it did theoretically mean that there were no obvious flaws.

I wanted to get to the bottom of it. I wanted to figure out whether bubble fusion was real. If it was, it could be the biggest science news to come around in a long time. I wrote back to Coontz. Of *course* I wanted to cover the story.

———

When Coontz first sent me the paper on February 5, he told me of an additional complication. The editors were afraid of an embargo break.

The embargo system is the dirty little secret of science journalism. Over the past few decades, science journalists entered into a compact with peer-reviewed journals like *Science, Nature,* and the *New England Journal of Medicine.* The journals provide copies of manuscripts to reporters a few days ahead of publication; these journalists, in return, agree not to tell the public about the manuscripts until the embargo expires, usually the evening before the peer-reviewed journal is published. Journalists who break the embargo, publishing ahead of the set time, are threatened with the loss of access to advance manuscripts, putting them at a great disadvantage with respect to their peers who abide by the rules. Nonetheless, some stories are so juicy that reporters can't resist; word inevitably leaks out before the embargo expires (often the fault of British newspapers, whose reporters are particularly jumpy). The embargo breaks, and it's a free-for-all.

Bubble fusion was obviously a juicy story, so the slightest word leaking to the press could trigger a media feeding frenzy. It was crucial to the editors that nothing be reported in the newspapers until the final version of the manuscript was ready. If the press started talking about the experiment before the paper was available, it could easily be a repeat of the cold-fusion disaster—science by press conference. It would not be fair to Taleyarkhan and his colleagues, who went through the peer-review process, to open them to accusations of subverting the system. Security had to be extraordinarily tight.

The paper was to be published on February 14. I was asked not to contact any scientists other than the authors until February 8, so that word of the paper wouldn't spread. It wasn't an extraordinary request, and I could certainly hold off on some of the phone calls for three days. But I had to contact Taleyarkhan right away. After I digested the paper, I sent him an e-mail to set up an interview; I also asked whether he was the same Taleyarkhan of the variable-speed bullets.

He was. He remembered the story I had written in 1999 and hinted darkly in his e-mail reply that the government might try to keep bubble fusion a secret, just as they had done with his earlier research. "Right after you did your story my project got classified. I hope something like this does not happen to this area. That would be a shame." In the rest of the note Taleyarkhan clearly showed that he thought he had made an important discovery. "This current area could have somewhat revolutionary and far-reaching consequences with very significant impacts on everyday life and a variety of disciplines (ranging from materials synthesis to medicine to food sterilization to counter-terrorism to power production and the like)." Power production. There it was. He hedged, and he put it last in his list, but it was there. Taleyarkhan thought he had found a path to fusion energy. This was going to be a big story, one way or another. Taleyarkhan and I made an appointment to speak on Friday, February 8.

On the evening of February 6, I learned that the bubble fusion article was on hold. Taleyarkhan's employer, Oak Ridge National Laboratory, was trying to apply the brakes. Gil Gilliland, an associate director of Oak Ridge, apparently called *Science* and complained that the paper had not yet passed Oak Ridge's internal review process, which had been ongoing since November. (This was an unusually long time to spend on a review.) Gilliland promised to get the review finished as soon as humanly possible, and the manuscript was rescheduled for publication on March 8. I was asked to hold off on the interviews until the paper was back on track. I didn't know it at the time, but the scene at Oak Ridge was getting ugly.

In fact, a battle was brewing. A number of physicists at the laboratory were extremely doubtful of the bubble fusion research, and their doubt triggered a flurry of activity behind the scenes. Shortly after Taleyarkhan submitted the manuscript—with Oak Ridge's permission—to *Science*, the skepticism in the lab began to mount. Lab officials apparently asked two other Oak Ridge scientists, Dan Shapira and Michael Saltmarsh, to repeat the bubble fusion experiment. Saltmarsh was a fusion scientist who had testified before Congress about the cold-fusion affair. Shapira studied exotic fusion reactions induced by high-energy beams of ions. Both had the expertise to find neutrons from bubble fusion—if those neutrons existed.

Calling for other scientists to repeat an experiment before publication was an extremely unusual step, and it likely struck Taleyarkhan as a vote of no confidence, but Oak Ridge insisted. The lab seemed determined to avoid becoming the center of another cold-fusion fiasco. So, with Taleyarkhan's assistance, Shapira and Saltmarsh set up an exact copy of the bubble fusion experiment, except for one detail: they used a bigger and better neutron detector. Not only was it physically larger (making it more sensitive, because more neutrons could strike it), but it also had more sophisticated electronics. Unlike Taleyarkhan's detector, it could tell the difference between neutrons and gamma rays.

When Shapira and Saltmarsh analyzed the data they had gathered, the results were damning. They found no sign of fusion, no evidence for neutron emission from the bubbling deuterated acetone. They did not try to verify Taleyarkhan's findings of tritium, but noted that if the tritium had been produced by fusion, the bubbling solution should have produced a million neutrons per second, and that level of activity should easily have been picked up by the neutron detector. According to their equipment, though, nothing was happening in the bubbling liquid, just the expected number of chirps caused by stray neutrons produced by cosmic rays and the like. (And since the team members were making bubbles by zapping the tank with neutrons, a heck of a lot of those particles were skittering about in the background.)

Oak Ridge was in a bind. They were about to look foolish. One of their researchers was about to publish what they considered a bad piece of research that would spark a second cold-fusion fiasco. And they were increasingly powerless to stop it. The lab had already given Taleyarkhan permission to seek publication, and *Science* had already accepted and reviewed the manuscript. Yet Oak Ridge seemed to have an experiment that blew the Taleyarkhan discovery out of the water. They were rapidly running out of options.

Scrapping the paper ceased to be a possibility the moment Taleyarkhan had sent the paper to the journal. However, Oak Ridge's objections had slowed publication by a few weeks. In that time, the lab moved fast to try to reduce the impending damage. Shapira and Saltmarsh quickly typed up their results in a short report and sent it over to *Science*, hoping the two papers would be published side by side. The negative report, if accepted, would at least force readers to cast a skeptical eye on the claims of bubble fusion; Oak Ridge wouldn't look quite so bad when other researchers poked holes in Taleyarkhan's work. Unfortunately, that wasn't an option, either. Any scientific manuscript in *Science* had to be peer reviewed, and the Shapira-Saltmarsh paper was no exception. There was no way that a new paper could be sent to reviewers, receive comments, and be revised in time to make the March 8 issue. (And it was becoming increasingly clear that holding beyond March 8 would be impossible; word of Taleyarkhan's paper was beginning to leak out.) Oak Ridge had no options left. The world would soon learn about bubble fusion, even though Shapira and Saltmarsh had shown that it was almost certainly a fiction.

I was back on the case on the afternoon of Wednesday, February 20. The *Science* editors had decided to go ahead and publish the article on March 8, and while formal approval had not yet come through from Oak Ridge, they had assurances that it would come shortly. (And it did.) I was given the green light to begin reporting again, but I was warned to tread carefully to avoid leaks. I immediately e-mailed Taleyarkhan again and set up an interview. That part was easy. The hard part was figuring whom else to talk to.

I needed to speak to outside researchers, people not in Taleyarkhan's research group. Only then would I get a reasonably objective opinion on the quality of the paper. At this stage, I couldn't show the manuscript to anyone who hadn't yet seen it; I couldn't be responsible for a leak this far ahead of publication. So I had to figure out who had already seen the paper—I had to find the paper's reviewers.

Nobody at *Science* would tell me who they were. The reviewers are kept confidential, even from the reporters who work for the same magazine. But I could guess. The Taleyarkhan paper crossed two fairly established disciplines, sonoluminescence and fusion. Just a few groups had been studying sonoluminescence for years. Lawrence Crum led one at the University of Washington, Seth Putterman led another at the University of California, Los Angeles (UCLA), and Ken Suslick ran a third at the University of Illinois. I was fairly certain that at least one of these scientists had been a reviewer. The fusion side of the paper was tougher. It was a bigger field, with many more researchers. I figured that the most likely candidates were those who knew about the nitty-gritty of neutron detection. If anyone would be able to bolster or tear down Taleyarkhan's work, it would be a neutron expert. In fact, if I were to pick reviewers for the manuscript, I would choose some of the physicists who had dissected the cold-fusion papers. They would certainly approach the paper with a skeptical eye, and if they were convinced, the paper would automatically get a huge amount of credibility.

I began making discreet inquiries.* On the sonoluminescence end, I called Crum and struck pay dirt. (As it turned out, all three of the big names in sonoluminescence—Crum, Putterman, and Suslick—had been reviewers.) I got the distinct impression that the sonoluminescence people were impressed by Taleyarkhan's technique, if a bit skeptical about his team's conclusions. Crum, at the very least, seemed particularly interested in Taleyarkhan's method of creating large bubbles with a beam of neutrons and thought it might open some new opportunities for research. ("I thought, doggone! I'm depressed I hadn't done that experiment," Crum told me. "It's a remarkable result, and I would like very much for this to be true.") So the sonoluminescence end of the experiment seemed relatively solid, at least from my limited reporting.

By that time, the editors had given me more detailed neutron data from Taleyarkhan's lab. The new information didn't assuage my doubts. I was no expert at interpreting such data, but they didn't look quite right. They were muddy; the shape of the peaks in the graph didn't appear the way I expected them to. These were just my gut instincts, but they emphasized my need to find a neutron expert.

When the time came for my interview with Taleyarkhan, I found him open and friendly. He was happy to tell me all about the research. I confirmed that he was, quite naturally, enthusiastic about the quality of the results—including the neutrons he was detecting. However, no matter how confident Taleyarkhan was, he was not going to be the person who could assure me about the quality of the research. I still had not found a neutron expert who had already seen the paper, especially

---

* It wasn't too hard to inquire about the paper without causing any leaks. Just the fact that I was from *Science* meant that reviewers knew I was likely calling about the Taleyarkhan paper, while those who had not heard of the manuscript would assume that I was calling about something else.

since I was still supposed to be very discreet. That problem was about to be made moot.

I had sensed that the editors at *Science* were getting increasingly tense. By the twenty-sixth, I had heard rumors in the building that somebody was trying to pressure the journal to reject the Taleyarkhan paper, and that *Science*'s editor in chief, Don Kennedy, was hopping mad. I didn't know anything more for certain until the morning of February 27, when Coontz passed me a cryptic note. He told me I had to call Princeton's Will Happer and IBM's Dick Garwin.

Happer and Garwin were legendary figures in the community. They were the big guns of fusion (and of nuclear weapons). Garwin had helped design Ivy Mike; Happer was a former head of the government's JASON panel. Both had been at the top of the scientific hierarchy for decades and had been involved in debunking cold fusion. They weren't reviewers of the manuscript—I was pretty certain of that—but clearly they had seen it. And they apparently had some very strong opinions that they expressed to Don Kennedy.*

I didn't know what Happer and Garwin said, but I knew that Kennedy was furious. He felt that outsiders were trying to disrupt the peer-review process and derail a paper that had already been accepted for publication. "There was certainly pressure from Oak Ridge to delay, if not to kill, the paper," Kennedy told me when I interviewed him. "I'm annoyed at the intervention, I'm annoyed at the assumptions that non-authors had the authority to exercise constraints on the publication and telling us we couldn't publish the paper—which they did."

I called Happer and began to piece together what was going on. (Garwin was in China at the time but I soon got his side of the story,

---

* Some scientists had a low opinion of Kennedy, a biologist and environmental scientist, long before he arrived at *Science*. While president of Stanford University in the early 1990s, he was pilloried by Congress and the press for an accounting scandal having to do with Stanford's research accounts. I think that much of the criticism of Kennedy is unfair, but he was forced out as president, and many in the scientific community blamed him for increased government regulations and scrutiny on their funding. A decade had passed since his resignation, but some in the scientific community still bore him a grudge.

too.) Someone—I never found out for sure who it was—had sent Garwin and Happer each a copy of the Taleyarkhan manuscript and a copy of the Shapira-Saltmarsh paper. Both then e-mailed Kennedy. Garwin was harsh and succinct:

> I understand there has been some discussion as to whether *Science*, having accepted the paper, should nevertheless not print it. I certainly don't want to enter into such a discussion with you.
>
> But I do want to tell you that I have read both papers carefully, and that I think the odds are extremely high that this "discovery" is simply error and incompetence.
>
> So I caution you to mute the natural enthusiasm of people on your team who want to publish the latest significant discovery.

Happer's note was longer and more involved, but he, too, urged Kennedy not to hype the results, likening it to cold fusion and polywater:*

> I am told that the paper in question, the most recent version of which has the title "Evidence for Nuclear Emissions During Acoustic Cavitation," will have a place on the cover of "Science" when it is published, and will be accompanied by additional laudatory editorials and sidebars. I may well be misinformed about this, but if I am not, I hope I can persuade you to exercise your authority as Editor in Chief to stop these plans....
>
> Giving the Taleyarkhan paper lots of publicity in "Science" would damage the credibility of the magazine and would do quite a bit of harm to our scientific community. I have seen several examples of similar episodes in my career, for example,

---

* Polywater is another notorious case of pathological science from the 1960s. Russian scientists claimed they had found a new form of water, one more viscous than ordinary water. Scientists studied the phenomenon for a while before deciding that the viscosity was caused by contamination.

polywater and cold fusion. Both episodes caused lasting damage
to science.

Happer strongly argued that the Taleyarkhan paper was in error, and
said that several other prominent fusion researchers (including the laser
fusion scientist John Nuckolls) had concluded that Taleyarkhan was flat-
out wrong. But at the same time, Happer insisted that he wasn't trying
to block publication. "I like *Science*; I'm a member of AAAS,* and I don't
want them to shoot themselves in the foot—or some other body part,"
Happer told me. "All I told him was, for God's sake don't put it on the
cover of *Science*." Garwin was a little more oblique about what *Science*
should do with the bubble fusion manuscript. "It would be unfortunate
if *Science* magazine were to take any position on its correctness," he said.

There's no question in my mind that Happer and Garwin would
have been happy if *Science* had pulled the Taleyarkhan paper. They were
convinced (as I was becoming) that bubble fusion was a fiction. However, I think that by the time they contacted Kennedy, they knew that
Science was likely to publish the paper. They just wanted to minimize
the damage to Oak Ridge and to the journal and, above all, to science.
Don't hype the results. Don't put it on the cover. Don't go out on a
limb for bubble fusion.

Kennedy, not unreasonably, interpreted the notes slightly differently. He saw it as a last-ditch attempt by Oak Ridge to quash publication. His response to Happer was measured, if indignant. *Science* editors
had taken exceptional care with the manuscript, he insisted, and the
review process had led to a firm recommendation to publish:

> Concern on the part of research managers from ORNL appeared
> late in the game, followed by some secondary measurements
> taken with a different detector which are claimed to show dif-
> ferences with respect to some of the results. That work has not
> been peer-reviewed....

* AAAS, the American Association for the Advancement of Science, is the scientific
  society that publishes *Science*.

I am now hearing from you and a few other distinguished physicists arguing that I should now block publication of a paper that has met and passed all our tests. Because I believe that the right way to resolve serious scientific differences is through repetition, peer review, and publication I plan to proceed. I have told the ORNL management that we will be happy to consider a manuscript by those who have a different interpretation.

It was getting more complicated by the minute, but now that the Taleyarkhan paper was circulating around the physics community, I was freed from the high level of secrecy that had hampered my investigations. I made dozens of phone calls to scientists around the country, asking not only about bubble fusion but also about what had happened with the review process at Oak Ridge. I knew about Shapira and Saltmarsh, so I contacted them. I also spoke to one of Oak Ridge's deputy directors, Lee Riedinger.

I was pretty sure Riedinger thought bubble fusion was garbage—but he never said so directly. "I'm confused," he told me, when I asked him whether he believed the Taleyarkhan experiment or the Shapira-Saltmarsh one. "There's an active dialogue back and forth about what could be wrong with either set of measurements." Riedinger seemed to be walking a very fine line. He was trying to temper enthusiasm for Taleyarkhan's results without publicly faulting his own employee's research. (And he went out of his way to compliment Taleyarkhan's abilities, adding that his work is "very novel and interesting.")

I found a fusion expert at Livermore, Mike Moran, who had also done sonoluminescence—with deuterated water, no less—and I asked him what he thought of the paper. "The paper's kind of a patchwork, technically, and each of the patches has a hole in it," he told me, and pointed out a number of damning flaws that I hadn't considered. Taleyarkhan's experiment seemed to be producing some tritium, and the tritium production disappeared when the sound-wave generator wasn't operating. Moran pointed out that this disappearance was a problem. A deuterated solution that had been irradiated by neutrons should show an increase in tritium levels whether or not the sound-wave generator—which collapsed the bubbles to ignite fusion—was working. Even without the fusion reactions, some of those neutrons would strike deuteriums and stick, creating tritium. In Taleyarkhan's experiment, this was not the case; the control experiment with deuterated acetone and no sound-wave generator showed no increase in tritium. "If he's really right, it should have shown up," Moran said. "It's an inconsistency in the data."

I was convinced. Taleyarkhan was wrong: bubble fusion was a fiction. And because of the spurious result, a scientific drama was playing out before my eyes. The officials at Oak Ridge felt that the Shapira-Saltmarsh paper was damning, and they were hoping to avoid embarrassment. Garwin and Happer were trying to prevent another cold-fusion controversy, and Kennedy was trying to preserve the integrity of the peer-review process. Rumors were flying, and they were getting nastier and more paranoid by the minute. Everybody was getting increasingly annoyed with everyone else.

I got a note from Happer on Friday, March 1. He relayed part of a message from an unnamed colleague (not from Princeton or Oak Ridge) who was telling people of rumors that I had been "calling around asking about activities surrounding the publication of the article, not the 'science' in the article." That message continued:

> I don't want to get anyone upset about this, but it does tend to verify the rumors we have heard about Don Kennedy, the current Science editor and former Stanford President, wanting to go after Princeton people for opposing the publication of the research paper in its original form.

The story had just gotten harder. The allegations were absurd—Kennedy had not influenced me at all, so I could hardly be his attack dog. However, it would be impossible to get honest opinions from physicists who believed that I was preparing a hatchet job at the behest of Don Kennedy. Later in the day, Happer sent a final note:

> Be careful what you write and remember:
>
> "The moving finger writes, and having writ moves on.
> Nor all thy piety, nor all thy wit,
> Can call it back to cancel half a line
> Nor all thy tears wash out a word of it."

I had been Omar Khayyamed.

Happer need not have reminded me. Of course I was acutely aware of the sensitivity of the situation. People were quite likely to get their first impression about bubble fusion from my article at the front of the magazine. I was skeptical, and I wanted that skepticism to come through, but at the same time I wanted to make sure that everybody's view was represented fairly: the researchers, the editors, and the skeptics. In my mind, the whole bubble fusion controversy was fueled by these three parties' tragic mutual incomprehension.

The bubble fusion story was about more than a scientific paper. It had become a story about the way science is done—and how the scientific peer-review process sometimes fails. The original ending to my piece grimly emphasized the point:

> Taleyarkhan and his team, in good faith, submitted their paper to peer review and passed. *Science* magazine subjected the paper to scrutiny in accordance with their procedures, and once the paper was accepted, refused to allow outsiders to influence their publication process. And outside scientists, worried about the quality of the paper, tried to prevent the embarrassment of a second cold fusion fiasco. All three parties had the best of intentions—and now the road ahead is paved.

This ending was scrapped in favor of a less-editorializing one. Yet even as I penned those words, the misunderstanding built. Bob Park, the tireless critic of pseudoscience (particularly cold fusion and its proponents, such as Thomas Valone, the patent examiner), featured bubble fusion in his weekly *What's New* newsletter, which is distributed on Friday afternoons:

BUBBLE FUSION: A COLLECTIVE GROAN CAN BE HEARD

A report out of Oak Ridge of d-d fusion events in collapsing bubbles formed by cavitation in deuterated acetone, is scheduled for publication in the March 8 issue of Science magazine....Although distinguished physicists, fearing a repeat of the cold fusion fiasco 13 years ago, advised against publication, the editor has apparently chosen not only to publish the work, but to do so with unusual fanfare, involving even the cover of Science. Perhaps Science magazine covets the vast readership of Infinite Energy magazine.

Infinite Energy, of course, was the cold-fusion activist Eugene Mallove's publication. Park was uncharacteristically wrong about the cover; bubble fusion was never going to be a cover story. However, his comment meant that the hitherto private controversy was about to become very public.

Over the weekend, the rumors flew, as did the press releases. *Science* distributed one as part of the weekly embargoed notification to journalists:

FUSION IN A FLASH? SCIENCE RESEARCHERS REPORT NUCLEAR EMISSIONS FROM TINY, SUPER-HOT COLLAPSING BUBBLES

The dramatic flashing implosion of tiny bubbles—in acetone containing deuterium atoms—produces tritium and nuclear emissions similar to emissions characteristic of nuclear fusion involving deuterium-deuterium reactions. This finding was reported in the 8 March issue of the peer-reviewed journal Science, published by the American Association for the Advancement of Science.

Shock wave simulations also indicate that temperatures inside the collapsing bubbles may reach up to 10 million degrees Kelvin, as hot as the center of the sun. Although the high temperatures and pressures within the bubbles would be sufficient to generate fusion, the overall results of the study only suggest, but do not confirm, nuclear fusion in the bubbles' collapse....

The experiment's entire apparatus is well within the bounds of "table-top physics," about "the size of three coffee cups stacked one on top of the other," says Taleyarkhan....

Currently, the level of neutron emissions with the characteristic fusion energy appears to be lower than would be expected from the tritium signals observed in the experiment. Further tests are needed to account for this discrepancy, and to verify the observed relations between the neutron emissions, tritium production, and bubble collapse.

If fusion is confirmed in further tests, these bubbles would still have a long way to go before they could be considered as a possible energy source with any commercial value, says Science co-author Richard T. Lahey Jr. of Rensselaer Polytechnic Institute. First of all, the bubble reaction would have to demonstrate net energy gain—that is, it should produce more energy than the energy needed to drive the reaction itself. Second, scientists would have to find a way to make the reaction perpetuate itself in a chain reaction, without constant input from a neutron source....

It was optimistic, but not outrageous. The biggest problem, in my mind, is that it did not mention the Shapira and Saltmarsh manuscript. That was a major oversight, though I am not sure whether the press office was even aware of the paper at the time.

Reporters who received Science's press package could get the Taleyarkhan paper, but the information was embargoed until 2 PM on Thursday, March 7. Only then would science reporters be allowed to print their stories.

There was also an Oak Ridge press release, but I didn't see it until a few days later. Its tone was a little more pessimistic:

PRELIMINARY EVIDENCE SUGGESTS POSSIBLE NUCLEAR EMISSIONS DURING EXPERIMENTS

Researchers at Oak Ridge National Laboratory, Rensselaer Polytechnic Institute and the Russian Academy of Sciences have reported results that suggest the possibility of nuclear reactions during the explosive collapse of bubbles in liquid, a process known as cavitation....

Experiments suggest the presence of small but statistically significant amounts of tritium above background resulting from cavitation experiments using chilled deuterated acetone. This tritium could result from the nuclear fusion of two deuterium nuclei. Tritium was not observed during cavitation of normal acetone, which does not contain deuterium.

Attempts to confirm these results by looking for the telltale neutron signature of the deuterium fusion reaction have yielded mixed results. While there are indications of neutron emission in the newly published results, subsequent experiments with a different detector system show no neutron production.

Theoretical estimates of the conditions in the collapsing bubbles are consistent with the possibility of nuclear fusion, under certain assumptions concerning the relevant hydrodynamics.

These results suggest the need for additional experiments, said ORNL's Lee Riedinger, deputy director for Science and Technology. In particular, the difference in the two sets of neutron measurements must be clarified. Additional tritium experiments would also allow a better understanding of the tritium observations.

Until confirmatory experiments are completed, a cautionary view is appropriate, according to Riedinger, who said, "The manuscript has been through external peer review, but the scientific record shows that tritium and neutron measurements at these levels are difficult, and one must do further tests before firm conclusions can be drawn."

Like Riedinger's comments to me, the Oak Ridge press release was as negative and cautious as it could possibly be without directly undermining Taleyarkhan. Everybody at Oak Ridge was carefully watching their words.

I was, too. Over the weekend, after a few last-minute e-mails, I put the finishing touches on the first draft of the article. I had been asked to send a copy to Don Kennedy, so I did. It was an unusual request. Kennedy, as the editor in chief of *Science*, had the final say over everything that appeared in the pages. However, like his predecessors, he was pretty hands-off, at least when it came to the news section. We reporters were reasonably insulated from the politics of the magazine.

On Monday, I heard back from Don Kennedy. He seemed a little annoyed by the tone of my piece. The references to cold fusion and the use of the word *tabletop* bothered him. I stood my ground, arguing that everyone was, rightly or wrongly, comparing bubble fusion to cold fusion and so we had to use the term. As for tabletop, Taleyarkhan had used the word, as had *Science*'s own press release. Kennedy immediately relented:

> I'm sorry if my cold-fusion allergy led to a slight grumble on my part. I don't see how you could have avoided that term, and table-top is certainly okay....I hope it was clear that although I might make a comment on a draft in such a situation, I am absolutely pledged to non-interference.

The term *cold fusion* was clearly driving the editors at *Science* to distraction. On the morning of Monday, March 4, *Science*'s press office admonished reporters who were even thinking of using the phrase:

> [W]e ask all Science Press Package registrants to note that the peer-reviewed Science paper describes reactions inside bubbles that reach temperatures as hot as the center of the sun—up to 10

million degrees Kelvin. Thus, descriptions of this work as "cold fusion" are grossly inaccurate. We wish to thank all journalists who are taking the time to read and understand this research, to convey accurate information to the public.

In the same notice, the press office told of the Shapira-Saltmarsh paper and offered to provide it to reporters, as well as a response by Taleyarkhan's team. (Oddly enough, Taleyarkhan's team looked through the raw data provided by Shapira and Saltmarsh and claimed to see evidence of neutrons that the two were ignoring.)

The situation had already reached a boiling point. By 1:30 PM on Monday, the embargo was blown. The press office lifted all restrictions on using the articles and begged, once more, that journalists write a balanced story.

"Here we go....Fasten your seatbelts," one editor told me. It was all over the Internet in seconds. My article was being made available to reporters as well, and I started getting phone calls from television producers inviting me to talk on the air about bubble fusion.*

The press coverage ran the gamut from optimistic and credulous ("Fusion 'Breakthrough' Heralds Cleaner Energy," trumpeted London's *Guardian*) to pessimistic and weary ("Here we go again; Table-top fusion," sighed the *Economist*). Most were in the middle. My impression was that television reporters (as usual) were more keen on bubble fusion than their print counterparts, but few went overboard. After an intense burst of interest for a week or so, the media frenzy began to calm down. But an undercurrent of bad feeling remained within the science community.

I knew that *Science* would be vulnerable to attack because of the bubble fusion paper, but I was surprised by the source of the most damning criticism. A week after the story broke, three of the reviewers

---

* I didn't appear on TV. When I got the first call, I contacted Don Kennedy, telling him that I'd be willing to talk about the science behind the papers and what people on all sides had told me, but that it would be inappropriate for me to represent *Science* magazine. I suggested that the editors should handle the television interviews. He agreed, and I was spared the television circuit.

of the paper—Putterman, Crum, and William Moss, a sonoluminescence theorist at Livermore—told the *Washington Post* that *Science* had published the Taleyarkhan paper over their objections:

> "I reviewed the paper twice, I rejected it twice," said William Moss, a physicist at the Lawrence Livermore National Laboratory in California.

> "I told Science you can't publish it because it's not right," said Lawrence Crum, a physicist with the Applied Physics Lab of the University of Washington at Seattle.

> "They say it was subject to stringent peer review, but does that mean it passed peer review?" asked Seth Putterman, a physicist with the University of California at Los Angeles, who also rejected the article.

Now this—this surprised me. I had been so busy looking for the neutrons that I didn't spend a lot of time with sonoluminescence physicists. And what I heard from them had been complimentary, if cautious. (After all, Crum had even used the word "doggone" when describing the beauty of Taleyarkhan's idea!) Shortly after the *Washington Post* story, Ken Suslick, too, chimed in. In the beginning of April, Putterman, Suslick, and Crum wrote a short criticism of the Taleyarkhan paper, arguing that it had been "unready for publication" and suffered from "substandard experimental techniques."

The public criticism was not coming from fusion scientists, but sonoluminescence people,* and from those I thought were reasonably supportive of Taleyarkhan's technique. When I had interviewed Crum, and he admitted that he was a reviewer, he seemed positive enough that it didn't occur to me to ask whether he suggested rejecting the paper. I had completely missed it. The three reviewers were also heaping criticism on *Science*'s review process. Later in the year, Putterman chal-

---

* Putterman had not yet published his work on piezoelectric fusion (see appendix), so I was unaware of his research in that area.

lenged *Science* to publish the positive reviews: "Somewhere out there is a positive report from someone," he told *Nature*, *Science* magazine's main rival. "*Science* should publish that report because then we'll see what kind of information they went on to overrule four negative reviewers." Of course, Don Kennedy refused. "We maintain our end of the confidentiality bargain about peer review, so I can't discuss the process specifically, except to say that the positive reviews outweighed the negative ones. Why else should we publish the paper?"

Did *Science* overrule the virulent objections of its reviewers and deliberately publish a bad story? Were the sonoluminescence people jumping on the anti-bubble-fusion bandwagon after it got hammered by Shapira, Saltmarsh, and other fusion scientists? Unfortunately, I don't know for sure.

Science is a peer-reviewed journal. But it is also a magazine. And magazines, especially those that run advertisements and classifieds, are always trying to boost their circulation. Peer-reviewed journals like to publish provocative and spectacular results in their pages to get extra attention. Sometimes this leads to bad science; even the best peer-reviewed journals occasionally publish substandard manuscripts in their pages. (*Nature*'s letters section, for example, is notorious for occasionally publishing attention-grabbing but dubious research.) The editor who received the bubble fusion manuscript might have been influenced by the spectacular nature of the claims, but at the same time, I don't think that he (or other editors at *Science*) were consciously gaming the peer-review system to accept a manuscript that its reviewers had clearly rejected.

Though I have never gotten my hands on them, I believe that the reviews of Taleyarkhan's paper were mixed—skeptical and admiring at the same time—and that there was enough innovation in the experiment's technique that the editor in charge felt that the reviews were sufficiently positive to merit publication. However, even today, I have to shrug my shoulders when asked what happened behind the scenes during Taleyarkhan's peer review. Frankly, when the bubble fusion frenzy diminished, I was relieved.

The story continued to simmer in the background. In July 2002, I covered an article by Suslick that appeared in *Nature*. Suslick and a colleague had used fluorescent dyes to measure the by-products of sonoluminescence in water and compared them to theorists' expectations. "They're saying, 'We understand what's going on inside the bubble,' and if this is what you believe the science is, you should be suspicious of the Taleyarkhan paper," Crum told me. (I wasn't able to reach Taleyarkhan for comment by deadline, unfortunately.) After my article ran, Putterman dropped me a note saying that I was taking the theoretical work too seriously. I didn't agree entirely, but the point was valid if a little odd. He seemed not to want to dilute the direct criticism of Taleyarkhan's *Science* experiment with less-direct criticism from a theoretical perspective.

Taleyarkhan had also been busy. Oak Ridge had included him on a team of scientists assembled to attempt an experiment that would end the controversy once and for all, but the Department of Energy refused to pony up the necessary money. In 2003, Taleyarkhan quit his post at Oak Ridge in favor of a named chair in Purdue University's School of Nuclear Engineering. In 2004, he published a paper in *Physical Review E*, a high-level peer-reviewed physics journal, that seemed to confirm his original findings. I didn't think it added much to the debate, so I didn't cover it, despite an overheated press release from Purdue:

EVIDENCE BUBBLES OVER TO SUPPORT TABLETOP NUCLEAR FUSION DEVICE

Researchers are reporting new evidence supporting their earlier discovery of an inexpensive "tabletop" device that uses sound waves to produce nuclear fusion reactions....

...Whereas data from the previous experiment had roughly a one in 100 chance of being attributed to some phenomena other than nuclear fusion, the new, more precise results represent more like a one in a trillion chance of being wrong, Taleyarkhan said.

The *New York Times* covered the paper, as did a few other outlets, but it didn't spark a huge amount of discussion, even though Crum told the *Times* that the new work was "much better" than what had appeared in *Science*. Nor did another confirming paper, written by Adam Butt and Yiban Xu, that appeared in a lesser journal, *Nuclear Engineering and Design*, raise any eyebrows. In this case, Purdue's press release stressed the independence of the new work:

PURDUE FINDINGS SUPPORT EARLIER NUCLEAR FUSION EXPERIMENTS

Researchers at Purdue University have new evidence support-ing earlier findings by other scientists who designed an inex-pensive "tabletop" device that uses sound waves to produce nuclear fusion reactions.

The technology, in theory, could lead to a new source of clean energy and a host of portable detectors and other ap-plications....

Xu and Butt now work in Taleyarkhan's lab, but all of the research on which the new paper is based was conducted be-fore they joined the lab, and the research began at Purdue be-fore Taleyarkhan had become a Purdue faculty member. The two researchers used an identical "carbon copy" of the origi-nal test chamber designed by Taleyarkhan, and they worked under the sponsorship and direction of Lefteri Tsoukalas, head of the School of Nuclear Engineering.

Taleyarkhan saw this as a great achievement, and in 2006 he wrote about how his results had been "independently confirmed." This claim, too, was about to be challenged.

In 2005, I resigned from *Science* magazine to become a professor of journalism at New York University. So when the bubble fusion affair exploded again, I was watching from the sidelines. It was a much more comfortable position.

———

The bubble fusion affair began very differently from the cold-fusion fi-asco. Unlike Pons and Fleischmann, Taleyarkhan and his team had not sought publicity until their research had been peer-reviewed by a major journal. Once *Science* stamped its imprimatur on the work, then the group could claim they were doing the right thing, at least according to the traditions of science—they were steering debate about the work into the scientific literature. Of course, there were questions about their competence as well as their conduct. For example, why hadn't they withdrawn their paper after the devastating counterexperiment by Shapira and Saltmarsh?

I suspect they felt that Oak Ridge was trying to undermine their work and rob them of a publication in a prestigious journal. They were unconvinced (and perhaps uncomprehending) of the importance of the Shapira-Saltmarsh paper. The embattled bubble fusion scientists began to get paranoid, too. When the story first broke, Taleyarkhan's co-author (and PhD thesis adviser), Richard Lahey, told the *Washington Post* that criticism of the paper was "political" and motivated by hot-fusion physicists who were trying to hold on to their big budgets. Throw on top of that their desire to patent the device and profit from it, and I doubt they ever seriously considered withdrawing the paper. In my opinion, not withdrawing, even despite the Shapira and Saltmarsh counterevidence, did not cross the line into scientific misconduct, though some anti-bubble-fusion scientists seemed to believe otherwise.

With bubble fusion, there were no moving peaks, no firm accusations of scientific fraud—at least at first. As time passed, though, the story of bubble fusion did begin to mirror the cold-fusion fiasco ever more closely. An embattled Taleyarkhan would soon find himself under investigation and accused of fraud.

The story began to stir again in 2005, in part because of a BBC science documentary. BBC's science show, *Horizon*, interviewed Taleyarkhan and commissioned Seth Putterman to redo the Taleyarkhan experiment on TV. The results were completely negative. How many neutrons had Putterman seen coming from a bubble fusion cell? None at all. When the program, entitled "An Experiment to Save the World," aired, Taleyarkhan looked foolish—even talking about his hopes for a Nobel. "Nuclear fusion is a major finding, some people think that it may be worthy of a Nobel Prize," he said on the show. "It would be nice if it were. But I don't, I don't keep dreaming about it just now, if it happens so be it." Apparently, as filming progressed, Taleyarkhan got increasingly suspicious, and he refused to help the Putterman team with the experiment.

DR RUSI TALEYARKHAN: I would help out anybody who I feel, who I felt comfortable with. I would, I would, but I have to be comfortable with that particular group.

INTERVIEWER: Why, why is that, because is it not just science?

DR RUSI TALEYARKHAN: I will not answer that question right now.

Despite Taleyarkhan's reservation, within a month of the show's air date he and Putterman (along with Suslick) yoked themselves together with a grant from the Pentagon. The Defense Advanced Research Projects Agency (DARPA) gave them more than $800,000 to try to replicate Taleyarkhan's results. It was about that time that the murmurs about Taleyarkhan's incompetence began turning to murmurs about scientific misconduct.

Lefteri Tsoukalas was one of the people at Purdue who helped recruit Taleyarkhan from Oak Ridge. He apparently began doubting Taleyarkhan's integrity early in 2004, when the bubble fusion researcher first started working on campus full-time. Tsoukalas and five of his colleagues had been trying to replicate Taleyarkhan's work without any success, and they hoped that Taleyarkhan could help them get the experiments running properly. But when Taleyarkhan arrived, Tsoukalas's team was put off by his increasingly bizarre behavior. According to an exposé in *Nature*, Taleyarkhan allegedly started observing positive results that nobody else could detect: "He said: 'Look, there's a peak,' but there was nothing to see," one lab member told the magazine. "I started questioning it." And then, in May 2004, apparently without warning, Taleyarkhan reportedly removed the bubble fusion equipment from the department's lab. Though Tsoukalas and his colleagues were upset, they told *Nature* that they didn't press the issue in the interest of faculty harmony. Then Taleyarkhan apparently argued against publication of the Tsoukalas group's negative results. Yet, shortly after the Xu and Butt paper came out in 2005, Taleyarkhan is said to have pressed Purdue to issue the laudatory press release.

Soon, accusations started flying in the press that the Xu-Butt paper was not as independent as Taleyarkhan insisted, and Taleyarkhan soon found himself formally accused of scientific misconduct. In March 2006, Purdue University began an investigation into his actions, and the Xu-Butt research took center stage. Tsoukalas and a colleague, Martin Lopez de Bertodano, claimed that the Xu-Butt paper was "nothing but a contrived and hurried attempt to stage the appearance of 'independent confirmation' of sonofusion claims."

Things got worse for Taleyarkhan by the day. More scientists joined the chorus crying fraud. Ken Suslick used the f-word in an interview with the *Los Angeles Times* when *Nature* first aired concerns about the Xu-Butt paper: "Presenting that as independent is fraud," Suslick told them. And scientists had found other reasons to be concerned about Taleyarkhan's conduct.

Earlier in 2006, a scientist in Putterman's research group, Brian Naranjo, argued that Taleyarkhan's data were consistent not with fusion reactions but with the radioactive decay of an element known as californium-252. (Taleyarkhan had been "negligent or jumped the gun or concocted data—one of those," Putterman told the *Los Angeles Times*.) In May, Taleyarkhan's group admitted that it had made an embarrassing error with a key piece of equipment; its detector was made of a material different from what had been reported. In December, *Nature* reported on allegations that "data from Xu's paper are apparently identical to separate data reported by Taleyarkhan." Taleyarkhan's work was looking incompetent at best, and fraudulent at worst.* Scientists were expressing their concerns to Purdue.

When Purdue wrapped up its investigation in December, after repeatedly being accused of foot-dragging, it initially kept the results secret. In February 2007 the university made them public.

---

* Putterman also leveled a charge about using some of the DARPA money for an experiment that was not supposed to be part of the grant.

Committee members had found a number of disturbing issues. They concluded that the Xu and Butt research was not independent after all. Taleyarkhan's involvement in the work was consistent with that of a coauthor. In fact, his contribution was arguably greater than Adam Butt's. Consequently, the committee deemed that when Taleyarkhan kept his name off the paper, he showed "a severe lack of judgment." Despite this, the committee concluded in December 2006 that Taleyarkhan did not intend to mislead the scientific community and recommended against further pursuit of the matter.

Nevertheless, it would be further pursued. The committee had only looked into the allegations of fraud leveled by Tsoukalas and Lopez de Bertodano; it did not address many of the other complaints against Taleyarkhan that had been streaming in from outside researchers, and this raised questions about Purdue's fraud-finding mechanisms. Congressman Brad Miller, the chair of the House Science and Technology Committee's Investigations and Oversight panel, got Congress involved. Purdue University (and Taleyarkhan) received taxpayer dollars. If Purdue was not properly investigating fraud in its ranks of scientists, then Congress had reason for concern.

In March, Miller requested copies of the investigation reports. The request letter was ominous. "Despite the University's statement that no misconduct had occurred, many disturbing questions remain about the scope and quality of the information," Miller wrote. When Miller's staff reviewed the documents, they concluded that the investigation had not been thorough, had failed to address the validity of Taleyarkhan's research, and had not even followed Purdue's internal guidelines for investigating allegations of scientific misconduct. Prodded by Miller, Purdue sheepishly began another inquiry.

When I spoke to Taleyarkhan in August 2007, he was downcast. "I am exhausted, Charles. I'm very happy to speak with a friend, or [someone] who used to be a friend," he told me. "It is devastating. I've got two children who go to Purdue, who are students. One day in the press they see their father so honored and the next day they see him vilified. So, it's tough." It was heartbreaking to see this man—who had been so genuinely excited about his discovery—brought so low.

After nearly a year of deliberations, in July 2008, Purdue's last inquiry panel finally released its findings. It concluded that, yes, Taleyarkhan had committed scientific misconduct. According to the panel's report, Taleyarkhan had deceived the scientific community by falsely claiming the Xu and Butt paper was independent confirmation of Taleyarkhan's original bubble fusion paper. Moreover, adding Adam Butt's name as a coauthor—when Butt had contributed little to the work—was deemed a deliberate attempt to fool scientific reviewers into thinking that the research was more solid than it actually was. (Earlier, a reviewer had complained that a version of the manuscript, which had only Xu as an author, "was apparently done by one person so that needed cross-checks and witnessing of results seem lacking." Adding Butt to the manuscript defused such objections.) In short, according to the Purdue findings, Taleyarkhan had deliberately misled the scientific community to cover up the shortcomings of his work. (Further, it didn't find convincing evidence that Taleyarkhan had faked any experiments or fudged data.)*

Bubble fusion, like cold fusion, was steadily driven to the fringe of science. Though bubble fusion started out at the core of establishment science, it ended as sordidly as the cold-fusion fiasco had. The scientific community moved quickly from mere skepticism to accusations of fraud. Like Pons and Fleischmann before him, Taleyarkhan became increasingly bitter and isolated. As his experimental evidence came crashing down around him, according to the Purdue findings, he apparently turned to scientific misconduct to cover the shortcomings of his work.

Taleyarkhan's reputation will likely never recover. In early 2002, Taleyarkhan was a distinguished engineer. Six years later, he was an outcast. He threw his career away chasing after the hope of unlimited fusion power—and after dreams of the Nobel Prize that would come from solving the world's energy problems.

---

* The panel also found that Taleyarkhan had reused data, that Taleyarkhan failed to acknowledge DARPA funding, and that he manipulated Purdue's press release about the Xu-Butt paper—but that these actions didn't constitute scientific misconduct.

CHAPTER 9

NOTHING LIKE THE SUN

They started at once, and went about among the Lotus-
eaters, who did them no hurt, but gave them to eat of the
lotus, which was so delicious that those who ate of it left
off caring about home, and did not even want to go back
and say what had happened to them, but were for staying
and munching lotus with Lotus-eaters without thinking
further of their return; nevertheless, though they wept
bitterly I forced them back to the ships and made them
fast under the benches.

—*THE ODYSSEY*, TRANSLATED BY SAMUEL BUTLER

Bubble fusion, like cold fusion, imploded under charges of
fraud and scientific misconduct. Though both methods still
have their supporters, both have now been swept to the fringes
of science. Without a spectacular reversal of fortune, that is
where they will remain.

Hot fusion now enjoys a monopoly. Mainstream scientists who
hope for fusion energy almost unanimously pin their hopes upon
inertial confinement fusion or magnetic fusion. Tabletop fusion and
muon-catalyzed fusion are not going to lead to energy production.

Bubble fusion and cold fusion were delusions. There are no other options.

Despite that distinction, since the 1990s fusion scientists have had to fight, with increasing desperation, to keep hot-fusion research alive. Now, two multibillion-dollar projects, one in California and one in France, will determine the future of fusion. If the projects succeed, they will allow nations around the world to free themselves from dependence on oil. But if they fail, it is possible that no amount of money will be sufficient to realize mankind's ambition to bottle the sun.

———

When it was conceived at the Geneva summit in 1985, the International Thermonuclear Experimental Reactor (ITER) quickly became magnetic fusion's best hope of achieving breakeven. Europe and Japan joined in the effort, and along with the Soviet Union and the United States, the four parties, together, agreed to pool their resources to build an enormous tokamak. It was to be the most ambitious international scientific project ever attempted.

Not only was ITER supposed to achieve breakeven; it was supposed to attain ignition and sustained burn. In theory, after the reaction was started, the plasma would heat itself and provide fusion energy as long as it had fresh fuel to consume; it would be like a furnace or a boiler, just needing periodic restoking while it provided continuous power. Though ITER would cost $10 billion, it would finally end the half measures of the individual countries' domestic fusion efforts. The cooperating world powers were confident that they would finally end the research phase of magnetic fusion. They would finally be building essentially a working reactor. After so many disappointments and failed promises, scientists from around the globe would usher in the era of fusion energy. It was a golden vision, but it wouldn't last.

A decade later, the USSR was no more. The United States was the only superpower left. Japan was in the throes of an economic crisis. Science budgets everywhere were declining, and in the United States the money available for fusion research was plummeting. ITER was in deep trouble.

In truth, ITER's trouble began at birth. Nobody had ever pulled off an international scientific project of such an enormous scale. Figuring out how to compress and ignite a plasma was only one of the problems that ITER proponents had to solve. Perhaps even trickier was the problem of distribution and containment of pork.

Politicians like to see direct benefits from the money they spend. This means they want cash to flow into the hands of the people who elect them. That is the law of pork-barrel politics—why Congress so regularly funds ridiculous multimillion-dollar projects like useless bridges in Alaska. New Mexico congressmen tend to be munificent to Los Alamos; California senators back Livermore; New Jersey politicians support Princeton. It's similar in other countries. Politicians always like to spend money to benefit their constituents.

ITER provided a porky dilemma. No matter where the ITER partners put the reactor, three of the four parties were going to have to spend their money on a machine in another country. Even if these partners managed to build much of the equipment domestically, cash (and talent) would have to flow overseas. This isn't good pork-barrel politics. The country where the reactor would be built would get the lion's share of the benefits of the project, and the others would see their money flow into the hands of a rival.

Even a decade after the Geneva meeting, nobody had agreed where ITER would be built. Rather than consolidating multiple international efforts into one big project, the need to distribute the pork among the parties led to just the opposite: duplication of effort. There were three centers—one in Germany, one in Japan, and one in the United States—devoted to designing the reactor.

Declining budgets made matters much worse. Fusion scientists in the United States had been making drastic cuts to their research program. They obliterated almost everything that wasn't part of a tokamak project; the nation put almost all its magnetic fusion eggs in the tokamak basket. Many fusion scientists thought that other configurations (including some new ones like "spheromaks") might lead to a working reactor faster than a tokamak would. In their view, cutting off research for these alternatives was shortsighted and premature. The tokamak shouldn't be the only game in town. Thus, they were against ITER. They didn't want to wager everything on a single enormous tokamak. Moreover, they weren't alone in their wariness of the international reactor. Even tokamak physicists felt threatened, because the domestic fusion program would have to be gutted in favor of the enormous international collaboration. The already stretched budgets would have to accommodate ITER. Congress would not provide additional funds for more big domestic experiments, and the existing ones would be quickly shut down to cut expenses. Laboratories like Princeton's would become superfluous without a major machine to experiment with. There would only be one big machine in the world, and it would likely be overseas.

Thus, by the mid-1990s, ITER had a large number of opponents: non-tokamak fusion scientists who resented the single-minded concentration on tokamaks, tokamak physicists who were afraid of having the domestic fusion program shipped overseas, and most important of all, politicians who saw taxpayer money flowing into the hands of other countries' governments. Everybody, in theory, liked the idea of a huge international fusion effort. In practice, though it was unpopular, and budgets were still in free fall.

By 1995, the magnetic fusion budget had been hovering around $350 million per year. The President's Committee of Advisors on Science and Technology (PCAST), an independent panel of experts that counseled the president on all matters scientific, gave Bill Clinton a grave warning about the fusion budget. At $320 million per year, the domestic program would be crippled, and ITER—as planned—would be too expensive to support; it would have to be renegotiated at a lower cost. A demonstration fusion power plant would be at least forty years away. If the budget dropped below $320 million, the consequences were almost too horrible to contemplate. The committee tried to envision a worthwhile fusion program with lower levels of funding but came to the following conclusion:

We find that this cannot be done. Reducing the U.S. fusion R&D program to such a level would leave room for nothing beyond the core program of theory and medium-scale experiments...no contribution to an international ignition experiment or materials test facility, no [new domestic tokamak], little exploitation of the remaining scientific potential of TFTR, and little sense of progress toward a fusion energy goal. With complete U.S. withdrawal, international fusion collaboration might well collapse—to the great detriment of the prospects for commercializing fusion energy as well as the prospects for future U.S. participation in major scientific and technological collaborations of other kinds.

When Congress passed the 1996 budget, magnetic fusion got about $240 million. It did not take long for things to unravel completely.

In the meantime, the projected costs for ITER were skyrocketing, and scientists raised new doubts about whether it would achieve ignition at all. Despite the rosy picture painted by the design team, some physicists predicted that new instabilities would cool the plasma faster than expected, meaning ITER would fail, just as generations of fusion machines had failed before it. If ITER was going to fail to achieve ignition and sustained burn, then, some physicists began to argue, domestic devices could fail just as well at half the price. The American scientists (as well as their Japanese counterparts, who were also cash strapped) started talking about scaling it back, making it into a less-ambitious experiment at a lower cost. ITER-Lite, as the plan was known, would only cost $5 billion. However, ITER-Lite would be unable to achieve ignition and sustained burn. It would be just another incremental improvement on existing devices.

Though ITER-Lite was cheaper, it would defeat the main benefit of pooling four countries' resources. No longer would the countries be leapfrogging over what domestic programs had been able to accomplish on their own. ITER-Lite would not be a great advance over previous designs. It would just be a massive, more expensive version of what everyone else had already built.

In late 1997, Japan asked for a three-year delay in construction. It was a terrible sign, and the designers scrambled to bring down ITER's costs. Physicists and engineers proposed various versions of ITER-Lite, but without the promise of ignition and sustained burn the troubled project was doomed. The United States decided it wanted out.

In 1998, the House Appropriations Committee noted angrily that "after ten years and a U.S. contribution of $345 million, the partnership has yet to select a site" for ITER, and slashed all funding for the project. (They even questioned whether a tokamak was the best way to achieve fusion energy.) In July, the United States allowed the ITER agreement to expire, refusing to sign an extension that the other parties had signed; in October, the U.S. pulled its scientists out of the ITER work site in Germany. ITER was dead, at least for the United States.

When ITER died, America's dream of fusion energy was officially deferred. Since the inception of the magnetic fusion effort in the United States, the government had considered it an "energy program"—Congress funded it in hopes of generating energy in the not-too-distant future. As ITER entered its death throes, the Office of Management and Budget changed magnetic fusion research into a "science program." This meant that the program's funding was no longer officially tied to the goal of building a fusion power plant. It was just pure research, science for science's sake. Consequently, it became a lower priority for Congress. An energy program was easy to drum up support for, but pure science was always iffier.

By the turn of the millennium, magnetic fusion was but a shadow of what it had been in the 1980s. The U.S. magnetic fusion budget stabilized at approximately $240 million, which was worth less every year as inflation nibbled away at the value of the dollar. The golden age of magnetic fusion was over in America.

Scientists in Europe, Russia, and Japan struggled to keep the ITER project alive without the United States' participation. They quickly decided that ITER, as originally envisioned, would be impossible to build. The three parties settled upon an ITER-Lite design. Gone was hope of ignition and sustained burn. Gone was the hope of a great leap toward fusion energy. And without the United States, even a drastically reduced ITER would be decades away.

In the meantime, fusion scientists had to make do with their increasingly obsolete tokamaks. They did their best to put a positive spin on a bad situation. Even as the original plans for ITER were dying, European and Japanese researchers finally claimed they had achieved the long-sought-after goal of breakeven. It was not as impressive as ignition and sustained burn, but if true, scientists had finally broken the fifty-year-old jinx and gotten more energy out of a controlled fusion reaction than they had put in.

In August 1996 and again in June 1998, researchers at Japan's JT-60 tokamak insisted that they had achieved "breakeven plasma conditions" and claimed their tokamak was producing 5 watts for every 4 that it consumed. A closer look showed that this wasn't quite what happened. JT-60 was using a plasma made of deuterium, so the fusion reactions in the plasma were entirely between deuterium and deuterium. These are less energetic than deuterium-tritium reactions. If you really want to get a magnetic fusion reactor producing lots of energy, you will use a mixture of deuterium and tritium as the fuel rather than pure deuterium. JT-60's "breakeven plasma conditions" did not really mean that the tokamak had reached breakeven. Instead, the JT-60 had reached pressures, temperatures, and confinement times that, according to calculations, *would* mean breakeven *if* researchers had used a deuterium-tritium mix rather than just deuterium as fuel. Every time JT-60 reached its "breakeven conditions," it was still consuming much more energy than it produced. So much for Japan's claim. What about Europe's?

JET, the big European tokamak, actually used deuterium-tritium mixtures in attempts to achieve breakeven. In September 1997, scientists loaded up a such a mixture into the reactor, heated it, compressed it, and…and what? What happened? It depends on whom you ask.

Some people insist that JET reached breakeven. Britain's Parliamentary Office on Science and Technology, for instance, states blandly in a pamphlet that "Breakeven was demonstrated at the JET experiment in the UK in 1997." This is a myth, just like the myth about JT-60. In truth, JET got 6 watts out for every 10 it put in. It was a record, and a remarkable achievement, but a net loss of 40 percent of energy is not the hallmark of a great power plant. Scientists would claim—after twiddling with the definition of the energy put into the system—that the loss was as little as 10 percent. This might be so, but it still wasn't breakeven; JET was losing energy, not making it.

National magnetic fusion programs are unable to achieve breakeven, let alone ignition and sustained burn. The national tokamaks like JET and JT-60 are reduced to setting lesser records: the highest temperature, the longest confinement, the highest pressure. However, these records are all but meaningless. Without getting beyond breakeven, the dream of a fusion reactor will remain out of reach. All the glowing press releases in the world won't turn an energy-loss machine into a working fusion reactor.

━━━━

Laser fusion scientists didn't suffer nearly as much in the 1990s as their magnetic fusion counterparts. As magnetic fusion budgets sank, laser fusion ones rose, because laser fusion scientists had a secret weapon: nuclear bombs.

Publicly, laser fusion scientists billed their experiments as a way to free the world from its energy problems. What John Emmett, a Livermore laser scientist, declared to *Time* magazine in 1985 was typical: "Once we crack the problem of fusion, we have an assured source of energy for as long as you want to think about it. It will cease to be a reason for war or an influence on foreign affairs." Emmett's optimistic vision was no different from what fusion researchers had been promising since the 1950s. Just like their magnetic fusion counterparts, laser fusion scientists had promised, again and again, unlimited, clean energy. Just like their magnetic fusion counterparts, laser fusion scientists had been disappointed again and again as instabilities and other problems demolished their overly optimistic predictions. Shiva had failed, and by the 1990s, so had Nova. Inertial confinement fusion's story was paralleling magnetic fusion's, down to the shattered dreams and broken promises.

Less loudly, though, scientists were pushing laser fusion for a completely different reason. They weren't really going after unlimited energy: they were pursuing laser fusion as a matter of national security. Without a working laser fusion facility, they argued, America's nuclear weapons arsenal would be in grave danger. Congress was sold. Even as magnetic fusion scientists were wringing their hands in the mid-1990s, their laser fusion brethren were rolling in money—thanks, in part, to the danger posed by the test ban.

On September 23, 1992, the United States detonated its last nuclear bomb, Julin Divider, before ceasing testing altogether. Throughout the 1990s, the world's nuclear powers were negotiating a permanent ban on nuclear testing. Though a few nations conducted a small number of such tests while the discussions went on, the United States held firm. No nuclear explosions.

Of course, nuclear testing was the way weapons designers evaluated their new warheads; no nuclear testing means no new types of nuclear warheads—more or less. There's some debate about whether the United States could manufacture slight variants on old weapons designs without resorting to underground detonations. However, it is certain that any sizable design change wouldn't be considered reliable until it was subjected to a full-scale nuclear test.

It's not a huge problem if the United States can't design new nuclear weapons; the ones on hand are sufficient for national security.* Instead, the test ban presented a more insidious problem. Without periodic nuclear testing, weaponeers argued, they could not be certain that the weapons in the nuclear stockpile would work. Nuclear bombs, like any other machines, decay over time. Their parts age and deteriorate. Since nuclear weapons use exotic radioactive materials, which undergo nuclear decay as well as physical decay, engineers don't have a firm understanding of how such a device ages. An engineer can mothball a tank or airplane and be certain that it will still function fifty or a hundred years from now. Not so for nuclear warheads. So, to assure the reliability of the nuclear stockpile, engineers would take aged weapons and detonate them to see how well they worked. With a test ban,

---

* A few hawks have pushed for new weapons designs. In the early 2000s, weaponeers were designing a controversial "bunker buster," the Robust Nuclear Earth Penetrator bomb, which probably would have needed testing before deployment.

though, scientists could no longer do this. Many weaponeers insisted there was no way to guarantee that the weapons in the nuclear stockpile would still work in ten or twenty or thirty years. So what was the government to do?

Enter the Science-Based Stockpile Stewardship program. Weapons scientists assured federal officials that with a set of high-tech experimental facilities they could ensure the reliability of the nation's arsenal. Some facilities would concentrate on the chemical explosives that set off the devices. Some would study how elements like plutonium and uranium respond to shocks. But the jewel in the stockpile stewardship's crown would be NIF, the National Ignition Facility at the Lawrence Livermore National Laboratory.

NIF is the successor to Nova. According to its designers, NIF, ten times more powerful than Nova, will zap a pellet of deuterium and tritium with 192 laser beams, pouring enough energy into the pellet to achieve breakeven. It will also ignite and have what is called *propagating burn*: at the center of the pellet, the fuel will begin fusing, and the energy from those fusions will heat the fuel and induce nearby nuclei to fuse. And of course, the fusion will produce more energy than the lasers put in. This is the same promise the designers made with Nova. And Shiva. But while Shiva cost $25 million and Nova cost about $200 million, in the early 1990s NIF was projected to cost more than $600 million. That number increased to more than $1 billion by the time the facility's construction started in 1997. That was just the beginning.

As late as June 1999, NIF managers swore to the Department of Energy that everything was peachy, that the project, which was scheduled to be finished in 2003, was on budget and on schedule. This was a lie. Within a few months, officials at Livermore had to admit to enormous problems and cost overruns. Some of the issues were simple oversights. The laser facility, for instance, had problems with dust settling on the laser glass. Dust motes would scatter the laser light and burst into flame, etching the glass. To fix this problem, NIF engineers had to start assembling laser components in clean rooms and tote them around by robotic trucks with superclean interiors—at enormous cost. That was just one of the issues that had to be solved with piles of money.

It was as if everything that could possibly go wrong with NIF was, in fact, going wrong. Some of the issues were minor annoyances: a brief delay in construction followed when workers found mammoth bones on the NIF site. Some were major: the glass supplier was having difficulty producing glass pure enough to use in the laser, forcing a revamp of the entire manufacturing process. Some were just bizarre. The head of NIF, Michael Campbell, was forced to resign in 1997 when officials discovered he had lied about earning a PhD from Princeton University.

Some problems were unexpected but easy to deal with, such as an issue with the capacitors, the devices that store the energy used to pump the laser glass. These devices were packed so full of energy that occasionally one would spontaneously vaporize. It would explode, spraying shrapnel around the room. Engineers solved the problem by putting a steel shield around the capacitors; when one exploded, flapper doors would open and the debris would spray toward the floor.

Some problems were more complex. For example, scientists had long since gone from infrared to green to ultraviolet light to reduce the disproportional heating of electrons compared with nuclei, but ultraviolet light at such high intensities was extremely nasty to optics. It would pit anything it came into contact with. The laser would damage itself every time it would fire. The solution was less than perfect: at NIF's full power, the optics will have to be replaced every fifty to one hundred shots or so, an extremely expensive prospect.

Furthermore, scientists were still struggling to deal with the Rayleigh-Taylor instability—the one that turns small imperfections on the surface of the fuel pellet into large mountains and deep valleys, destroying any hope of compressing the fuel to the point of ignition. Not only did scientists have to zap the target very carefully—so that the energy shining on the target was the same intensity on every part of the pellet—they also had to ensure that the pellet was extremely smooth. Even tiny imperfections on its surface would quickly grow and disrupt the collapsing plasma. To have any hope of achieving ignition, NIF's target pellets—about a millimeter in size—cannot have bumps bigger than fifty nanometers high. It's a tough task to manufacture such an object and fill it with fuel. Plastics, such as polystyrene, are relatively easy to produce with the required smoothness, but they don't implode very well when struck with light. Beryllium metal implodes nicely, but it's hard to make a metal sphere with the required smoothness. It was a really difficult problem that wasn't getting any easier as NIF scientists worked on it.

The cost of the star-crossed project ballooned from about $1 billion to more than $4 billion; the completion date slipped from 2003 to 2008. Worst of all, even if everything worked perfectly, even if NIF's lasers delivered the right power on target, nobody knew whether the pellet would ignite and burn. As early as the mid-1990s, outside reviewers, such as the JASON panel of scientists, warned that it was quite unlikely that NIF would achieve breakeven as easily as advertised. The prospects for breakeven grew worse as time passed. By 2000, NIF officials, if pressed, might say that the laser had a fifty-fifty shot of achieving ignition. NIF critics, on the other hand, were much less kind. "From my point of view, the chance that [NIF] reaches ignition is zero," said Leo Mascheroni, one of NIF's main detractors. "Not 1%. Those who say 5% are just being generous to be polite." The truth is probably somewhere in between, but nobody will know for sure until NIF starts doing full-scale experiments with all 192 beams.

If NIF fails to ignite its pellets, and if it fails to reach breakeven, laser fusion experiments will still be absorbing energy rather than producing it; the dream of fusion energy will be just as far away as before.* Furthermore, analysts argued, NIF wouldn't be terribly useful for stockpile stewardship without achieving breakeven. And NIF's contribution to stockpile stewardship is crucial for...what, exactly? It's hard to say for sure. Assume that NIF achieves ignition. For a brief moment, it com-

---

* In fact, many fusion scientists believe that even if NIF succeeds, it won't be a major advance on the path to fusion energy. NIF's lasers aren't the most promising candidate for fusion reactors. For one thing, they have to cool a long time between shots. Solid-state lasers, with their faster repeat rates and higher efficiencies, seem a more appropriate choice. Some work is also being done using two sets of lasers, one set to heat and one set to compress the plasma. NIF won't help much with this research either.

presses, confines, and heats a plasma so that it fuses, the fusion reaction spreads, and it produces more energy than it consumes. How does that translate into assuring the integrity of America's nuclear stockpile?

At first glance, it is not obvious how it would contribute at all. Most of the problems with aging weapons involve the decay of the pluto- nium "pits" that start the reaction going. Will the pits work? Are they safe? Can you remanufacture old pits or must you rebuild them from scratch? These issues are relevant only to a bomb's primary stage, the stage powered by fission, not fusion (except for the slight boost given by the injection of a little fusion fuel at the center of the bomb). The fusion happens in the bomb's secondary stage, and there doesn't seem to be nearly as much concern about aging problems with a bomb's sec- ondary. If the primary is where most of the problems are, what good does it do to study fusion reactions at NIF? NIF's results would seem to apply mostly to the secondary, not the primary.

Since so much about weapons work is classified, it is hard to see precisely what problems NIF is intended to solve. But some of the peo- ple in the know say that NIF has a point. The "JASONs," for example, argue that NIF does help maintain the stockpile—but not right away. NIF will contribute to science-based stockpile stewardship, the panel wrote in 1996, "but its contribution is almost exclusively to the long- term tasks, not to immediate needs associated with short-term tasks." That is, NIF will help eventually, but it is not terribly useful in the short term.

What are those long-term tasks? Two years earlier, the JASON panel was a little more explicit. NIF would help a bit with understanding what happens when tritium in a primary's booster decays. (However, since tritium has a half-life of only twelve years, it stands to reason that weapons designers periodically must replace old tritium in weapons with fresh tritium. This is probably routine by now.) NIF will also help scientists understand the underlying physics and "benchmark" the computer codes—like LASNEX—that simulate imploding and fusing plasma. (But why is this important if you are not designing new weap- ons? The ones in the stockpile already presumably work just fine, so you presumably don't need a finer understanding of plasma physics to maintain them.) The JASON members have access to classified information, but even so, their justifications for NIF seem a little thin—at first. And then JASON lists one more contribution that NIF makes to stockpile stewardship: "NIF will contribute to training and retaining expertise in weapons science and engineering, thereby permitting responsible stewardship without further underground tests." That's the main reason for NIF.

With the moratorium in place, nuclear tests are at an end. New scientists entering the program will never have a chance to design a bomb and test it. They will never have a chance to study a live nuclear explosion. All they have left are computer simulations and experiments that mimic one part of a nuclear explosion. NIF would be the only facility that mimics the explosion of a secondary; it would give young scientists a chance to study secondary physics without ever seeing a nuclear test. And that's the point of NIF. NIF is essentially a training ground for weapons scientists. As old ones retire and new ones grow up without ever having seen a nuclear test, NIF is a way to give them some level of experience so that America doesn't lose its nuclear expertise.

NIF isn't truly about energy. It is not about keeping our stockpile safe, at least not directly. It is about keeping the United States' weapons community going in the absence of nuclear tests. However, it is contributing next to nothing to the stockpile stewardship program at the moment, and the program is heading toward a crisis. Weaponeers are complaining that the United States is increasingly unable to vouch for its nuclear arsenal, and the government seems to be slowly slouching toward a resumption of nuclear detonations.

A number of ominous signs suggest that nuclear testing might begin again before too long. The debate in the early 2000s about the new Robust Nuclear Earth Penetrator warhead was an indication that the government was thinking beyond the test ban; before deploying the weapon, it almost certainly would need a test. Even though Congress strangled that program, it has blessed the Department of Energy's campaign to design yet another warhead. The Reliable Replacement Warhead (RRW), as it is called, is supposed to obviate the need for nuclear testing because it would be a hardier device less susceptible to aging. It would be able to assure the reliability of the nation's nuclear arsenal for decades without nuclear tests. The only problem is that the RRW would probably require a few nuclear tests before anyone was convinced of its reliability in the first place. It's a paradox: to maintain the nuclear test ban, the United States might have to resume testing.

A debate is also ongoing about shortening the time it will take to prepare the Nevada nuclear test site for a resumption of underground tests. President George W. Bush tried to make the site ready to resume testing within eighteen months, rather than maintain the previous twenty-four-month lead time. But going to a higher level of readiness announced to the world that the nation was moving toward ending the moratorium, and this could hamstring American attempts to stem the proliferation of nuclear weapons around the world. Year after year, the president put money for eighteen-month readiness in the budget; year after year, Congress took it out. Even without the cash, though, the National Nuclear Security Administration, the organization inside the Department of Energy responsible for nuclear weapons, lists eighteen-month test-site readiness as an integral part of the stockpile stewardship program.

The stockpile stewardship program will soon reach a crisis point. Will the federal government be able to assure the reliability of the stockpile without testing nuclear weapons as the program originally promised? Or will it fail, forcing a resumption of testing, breaching the moratorium in place for over a decade? The move toward renewed nuclear testing is happening now, and NIF, if it helps with stockpile stewardship at all, will do so indirectly and in the distant future. The nontesting regime might well be in tatters by the time scientists get any benefit from the multibillion-dollar machine supposedly designed to uphold it.

NIF is the state of the art in laser fusion, yet it is a deeply troubled project. It is vastly more expensive than originally projected. Even if it works perfectly, it won't keep the country's nuclear arsenal working or the nontesting policy alive. For a decade, experts have questioned whether it would be sufficiently powerful to achieve ignition and break-even—and if the history of laser fusion is any guide, NIF, like Nova, will fail to reach its goal. Yet NIF marches on. Laser fusion scientists won't give up their decades-old dream to put a star in a bottle. And if they fail, as it appears they will, after spending more than $4 billion, there is little hope that they can sucker the government into building yet another bigger and better laser machine.

———

In 2002, five years after the United States abruptly left the ITER project, fusion scientists were about to get a serious case of déjà vu.

The American departure shook the ITER collaboration—and branded the United States as an unreliable partner when it came to international science—but the project limped along. Russia, Europe, and Japan continued designing an international fusion reactor. The plans they came up with were much less ambitious than the original ITER. The plasma in the reactor would span twelve meters rather than sixteen meters. It would not achieve ignition and sustained burn—the plasma would never be fusing enough to keep itself warm—but if all went well, the reactor would be able to keep a plasma confined for up to an hour and produce ten times as much power as it consumed. (It would finally achieve breakeven—for real, this time.) It would cost half as much as the original ITER: $5 billion, rather than $10 billion.*

The American magnetic fusion program, in the meantime, was in ruins. There was no big domestic tokamak, just a few lesser ones in Boston and in San Diego. The big domestic tokamak, TFTR, had been

———

* Estimating costs of big projects conjures up all sorts of monkey business. As of 2003, the cost of construction was estimated to be about $5 billion. Inflation raises that cost by about 3 percent per year, but ignore that for the moment; 3 percent is small compared to some other factors that need to be considered. Major U.S. cutting-edge projects require a contingency fund—extra money to deal with unexpected problems that inevitably crop up. This contingency should be 20 percent or more of the construction cost, yet the $5 billion price tag doesn't include any contingency. So tack on $1 billion right there. That's just construction costs, assuming everything goes reasonably well. Operating the reactor and decommissioning it once it is done will cost at least as much as construction, so the total reaches $12 billion at a minimum.

shut down in 1997 to make room for ITER. Princeton, once home of the $100 million giants, was reduced to working on a tiny, $25 million spherical torus. Plans existed for larger machines, such as billion-dollar tokamaks, but they were just dreams; there was no chance they would be built. The United States was rapidly retreating from the cutting edge of magnetic fusion. Instead of getting a robust domestic program along with an enormous international reactor, American fusion scientists had neither. By 2002, with slim pickings at home, those scientists began to eye the slimmed-down ITER project, argued that many of the design flaws of the original machine had been fixed, and asked to rejoin the collaboration. At a cost of only about $1 billion, they argued, the United States could become an ITER partner again. The request worked its way up the food chain—from the scientists to a fusion advisory panel, to the head of the Department of Energy's Office of Science, to the secretary of energy, to the president. The answer was yes.

In early 2003, President Bush announced that the United States was back in the collaboration. The Americans would rejoin ITER.*

Even though the machine's design had been revamped and the collaboration had expanded—China, South Korea, and Canada had joined in—the same problems that haunted the first incarnation of ITER remained. For one thing the partners were still fighting over where the machine would be built.

Japan and Europe were the main contenders. Each attacked the other's proposal. Japan complained that the proposed European site in the south of France was too far from a port. The French argued that the Japanese site was prone to earthquakes. Most scientists in the United States understandably seemed to prefer a laboratory a short drive from the French Riviera to one near a dismal brackish lake in the north of

---

* The following year, Bush made NASA's primary goal a return to the moon, in part because it is home to "abundant resources" that can be exploited by humans. Other than water (which does not exist in great quantities on the moon), the main resource is helium-3, which can be fused with itself in a reactor to produce energy without creating many neutrons. The lack of any such device doesn't seem to trouble lunar-mission advocates, but it did make ITER, indirectly, a justification for an enormously expensive space program.

Japan, but the United States officially backed the Japanese site. Some Europeans hinted, darkly, that American support of Japan over France was political payback for France's criticism of the Iraq war. The Japanese accused the Europeans of circulating a nasty anonymous memo to the ITER parties that faulted the Japanese choice of site. China and Russia backed France. Canada pulled out of the collaboration entirely. Europe threatened to do so as well. In early 2005, more than three years after the United States had reentered the collaboration, ITER was deadlocked and on the brink of unraveling once again.

Back at the Capitol, Congress once again was getting very annoyed at the delay—and another old debate reopened. American fusion scientists started bickering about whether it was wise to decimate the domestic fusion program to fund an international reactor. The Department of Energy slashed its domestic programs to finance ITER; Congress restored the domestic funds and threatened to completely cut off money for the international reactor. ITER was about to collapse entirely.

Luckily for ITER's backers, the Japanese blinked just in time. Japan agreed that the French would host the reactor, but in return Europe would pay half the reactor's cost and would use Japanese companies for many of its manufacturing contracts. Furthermore, Japan would get to host a $600 million facility devoted to researching advanced materials for fusion reactors, materials that could withstand the intense heat and radiation inside a tokamak as well as reduce the amount of radioactive waste when the reactor vessel needed to be replaced. The debate was over. ITER would be sited in Cadarache, France. The American government, for its part, managed to find a way to fund its share: the fusion budget was increased to support ITER as well as the (modest) domestic program. India joined the collaboration. Everything seemed to be hunky-dory again.

On November 21, 2006, representatives of the seven ITER partner states signed the formal agreement. Everybody took the opportunity to wax poetic about what fusion power meant for the future. French president Jacques Chirac bubbled about ITER as a "hand held out to future generations":

The ambition is huge! To control nuclear fusion. To control the tremendous amount of energy generated at one hundred million degrees and to design sufficiently resistant materials for the purpose. To produce as much energy from a litre of seawater as a litre of oil or a kilo of coal.

It is a glorious vision. Unlimited energy—a tiny star bottled in a magnetic jar—would liberate mankind from the fear of global warming and from the impending energy crisis.

If ITER fails, it will probably mean the end of tokamaks. The likelihood of using magnets to confine and heat a plasma would seem slimmer than ever. However, there's no reason to assume that ITER, like generations of machines before it, will be a disappointment. If nothing goes wrong, ITER will begin experiments in 2018 or so.* And if ITER works as planned when scientists turn it on, it will light the way to a fusion reactor. If, miraculously, no more instabilities crop up that prevent scientists from bottling their plasma, fusion energy will be within reach. Scientists would then build a demonstration fusion power plant that would begin operations in 2035 or 2040. After five decades of broken promises, lies, delusions, and self-deception, it will finally be true. Fusion energy will be thirty years away.

---

* As this book went to press, things were going wrong on the U.S. side once again. Congress slashed funding for ITER, and while the president attempted to restore those funds, Congress was likely to cut them yet again. The other ITER partners may well have to forge ahead without American support.

CHAPTER 10

THE SCIENCE OF
WISHFUL THINKING

When one turns to the magnificent edifice of the physical sciences, and sees how it was reared; what thousands of disinterested moral lives of men lie buried in its mere foundations; what patience and postponement, what choking down of preference, what submission to the icy laws of outer fact are wrought into its very stones and mortar; how absolutely impersonal it stands in its vast augustness,—then how besotted and contemptible seems every little sentimentalist who comes blowing his voluntary smoke-wreaths, and pretending to decide things from out of his private dream!

—WILLIAM JAMES, "THE WILL TO BELIEVE"

We see what we want to see. That is why science was invented.

Science is little more than a method of tearing away notions that are not supported by cold, hard data. It forces us to discard ideas that we cherish. It eliminates some of our hopes, some of our dreams, and some of our wishes. This is why science can be so soul crushing to even its most devoted adherents.

Every scientist, at least on some level, has a vision of the way nature should behave. Every scientist, at least on some level, is wrong. And that means that scientists, sometimes subtly and sometimes unsubtly, occasionally try to wrestle the scientific narrative in the wrong direction. Like the mythmakers of old, they try to craft nature in their image.

The true power of science comes from its ability to withstand the wishful thinking of the humans who craft its stories. Individual scientists err. They deceive themselves—and they can deceive others. They might even lie or cheat in an attempt to win fame or glory or immortality. But the whole point of the scientific method is to try to insulate the scientific story from the whims and frailties of the scientists who write it.

The mechanisms of science are, essentially, protection against wishful thinking. This protection takes many forms, but the strongest come from the scientific community itself. Published scientific research is peer reviewed and vetted by rivals to ensure that its authors have made no obvious mistakes. The scientific community demands that experiments be repeatable, and if any question arises about the validity of an important experiment, scientists will clamor to have a second group verify the result with a different piece of equipment. And if there's a hint of incompetence or fraud, the community will howl for the blood of the malefactors. It can be brutal, but this is the way science protects itself from the dishonesty, the stupidity, or the human failures of an individual scientist. This is what makes science seem so inhuman. The scientific method has no sympathy for wishful thinking.

This can be hard on even the most brilliant scientists. As they practice their craft, they are forced to renounce some of their beliefs, no matter how deeply held they might be. If they err—as they almost certainly will—they must admit that they have deceived themselves. They have to do it publicly and without regard for their fragile human egos. They must eviscerate themselves on the altar of science. At least, that's what their peers expect.

———

For Andrew Lyne, an astronomer at the Jodrell Bank Observatory in England, the day of reckoning came in January 1992. Standing in front of a roomful of physicists and astronomers, Lyne was steeling himself, preparing to make an announcement that could destroy him. "It was the thing that one fears more than anything else in one's scientific life, and it was happening," Lyne said. "I certainly at the time thought that it was the end of my career."

Lyne was a radio astronomer, an expert in detecting and interpreting radio waves spewed out by stars and galaxies. In the early 1990s, his attention was drawn to a collapsed star known as a pulsar. These pulsars shine like cosmic lighthouses, emitting beams of radio waves as they spin. An earthbound observer like Lyne sees these pulsars blinking on and off with a clocklike regularity. But Lyne noticed that one pulsar was not blinking quite so regularly; it seemed to speed up and slow down. It was almost as if the pulsar was being tugged about by an unseen object, an invisible massive body orbiting the pulsar and pulling it out of its regular rhythm. He and his team had spotted what appeared to be a planet circling a foreign star.

Lyne was ecstatic. It would be the first detection of a world outside our solar system, a truly alien planet. It was something that astronomers had been looking for, in vain, for decades. This discovery would inscribe Lyne's name among the immortals of astronomy. Barely able to contain his excitement, Lyne submitted a paper to *Nature*.

The manuscript contained at least one significant issue. The planet seemed to orbit the pulsar once every 365 days, the same amount of time it takes the Earth to orbit the sun. It would be a pretty stunning coincidence if true, and to some astronomers it suggested something was wrong with Lyne's measurements. Perhaps he was failing to take the Earth's motion around the sun into account. It was a big warning sign, but Lyne was confident about his observations. "We did all sorts of tests on the data and tried to think of all the possible ways we might be making a mistake." They couldn't find an error. They were truly convinced: they were seeing an extrasolar planet. The reviewers at *Nature* were apparently convinced, too. It seemed to be a momentous discovery.

When the *Nature* paper came out, the astronomical community went wild. Lyne was showered with congratulations. The president of the American Astronomical Society immediately called a special session at the society's annual meeting to discuss the discovery. Lyne would be the guest of honor. Then disaster struck. "Ten or twelve days before I was due to give that talk, I discovered the error," Lyne said. It was a subtle one. His team had used the wrong piece of software to correct for the Earth's motion. With one of the dozens of pulsars they had been observing, they forgot to make a key change in the computer code. This minute error manifested as a tiny glitch in the pulsar's timing, a glitch that exactly mimicked the tug of an extrasolar planet orbiting the pulsar every 365 days.

The alien planet was a complete fiction. It vanished as soon as Lyne's team corrected the program. Less than a week before Lyne had to address his fellow astronomers—luminaries who had called a special session—the discovery dissolved into dust.

When Lyne took to the stage, he was petrified. "It was a large audience of extremely eminent astronomers and scientists," he said. However, he had decided what to do. Instead of telling everyone about the discovery of the extrasolar planet as originally planned, he told the gathered audience, in great detail, how he and his team had deceived themselves by failing to check their software properly. It was humiliating. Yet, at the end of his presentation, the audience broke out into a long, loud round of applause. Lyne was shocked. "Here I was, with the biggest blunder of my life and..." Lyne paused, gathering himself. "But I think that many people have nearly done such things themselves."

This is the way science is supposed to work. When a scientist discovers that he has erred, that he had deceived himself, he gives the scientific community a full and detailed report about his folly. The scientist abases himself, science rids itself of the erroneous notion, and the march of research continues on. However, reality isn't always so clean. Sometimes, other experimentalists join a scientist in self-deception; this makes it much harder to correct an error. It is also difficult when ego gets involved, as it often does. Lyne was lucky; he found his error himself. It's much harder to come clean when other scientists—your rivals—find your errors for you.

There are those who make a different decision. Many scientists, forced to stand on the edge of the abyss, gather their strength and leap. The annals of science are littered with the names of once-celebrated scientists whose wishful thinking forced them to jump into the fringe. If their pet theories become immune to contrary evidence, if their logic resists any criticism, if their peers suspect that they have fudged results, they are expelled from the scientific community. Usually this process takes years. With fusion, it can take just weeks.

Pons and Fleischmann were at the brink days after they went public. Almost immediately, Fleischmann in England and Pons in Utah discovered that their peak was in the wrong place—the gamma rays they thought they were detecting didn't have the right energy. They had to make a decision: retreat or press on despite the damaging evidence.

Taleyarkhan's group was nearing the brink even before their paper was published. At Oak Ridge, scientists had replicated the experiment with better neutron detectors and found nothing. It was a devastating blow. They had to make a decision: retreat or press on despite the damaging evidence.

The Taleyarkhan decision, at least at first, was more defensible than Pons and Fleischmann's. But in the end, they all wound up leaping into the void. Almost as soon as the researchers announced their results, accusations and investigations sent them to the fringe. The scientists of cold fusion and bubble fusion will never rejoin the ranks of the mainstream.

Every scientific field has its scandals and its renegades. There are biologists who dwell on the fringe, just as there are materials scientists, physicists, chemists, and geologists. But there's something about fusion that is a little different—the power of the dream of unlimited fusion energy that makes generation after generation of scientists deceive themselves.

The wishful thinking about fusion extends far beyond a handful of shunned individuals like Pons and Fleischmann, Taleyarkhan, and Perón's Ronald Richter. Individuals like these flare brightly and are quickly extinguished. They become the source of dark rumors and conspiracy theories, but they do superficial damage once they are excluded from the scientific community.

The real danger of wishful thinking comes not from these individuals but from the wishful thinking at the very core of the scientific community. This, and not the threat from a handful of renegades, is what makes the dream of fusion energy so dangerous to science.

The community seems to be in thrall to a collective delusion. Since the early 1950s, physicists have convinced themselves that fusion energy is nearly within their grasp. The perennially overoptimistic Edward Teller thought that within a few years, hydrogen bombs would carve canals, propel spacecraft, and generate almost unlimited amounts of energy. Lyman Spitzer thought powerful magnetic fields would create an artificial star within a decade. The ZETA team thought they had achieved fusion in 1958, freeing the planet from its dependence on fossil fuels. Laser fusion scientists thought that Shiva would produce energy, and that Nova would produce energy. Wrong, wrong, wrong. The history of fusion energy remained a series of failures.

Even if scientists finally change their luck, even if NIF breaks even and ITER manages to get a plasma burning for minutes at a time, both machines are still far from becoming working fusion reactors. NIF's design, particularly its slow lasers that need to cool for hours between shots, suggests that researchers will have to move to an entirely different type of laser system to have any hope of a practical energy source. ITER will never achieve ignition and sustained burn, the hallmark of a successful magnetic fusion reactor.

It is entirely possible that after billions of dollars and decades of research, fusion scientists will take the experimental results from ITER and turn them into a design for a viable fusion reactor. No physical law stands against it, after all. But if history is any guide, a long, long road lies ahead before physicists will be able to tame fusion reactions in a bottle.

Once they succeed, will it mean anything? Though aficionados are quick to tout fusion energy as the clean, unlimited energy source of the future, it is unlikely to be terribly clean or even terribly practical. The radioactive waste it generates is somewhat easier to deal with than the corresponding waste from a fission power plant, but it is a problem nonetheless. Fusion is also expensive. ITER is likely to cost $15 billion or more to build and run, and it won't ever be a practical reactor. Even with mass production, each fusion power plant will probably cost many billions of dollars. So long as there are other energy sources available, fusion is unlikely to make a huge dent in humanity's energy needs.

A better candidate, despite its unpopularity, is fission. Compare it to fusion. Both have a waste problem. Fission's is more severe, but not by much, at least in the near future. Fission plants are expensive, but they are likely to be considerably cheaper than their fusion counterparts. Fission plants are more dangerous than fusion plants (the fission reaction can get out of control, and a fusion reaction almost certainly won't), and malefactors can process spent fuel rods to get materials for atom bombs. However, new designs (such as pebble-bed reactors) reduce the risks dramatically. Fission plants don't have an unlimited source of fuel, but they do have enough for a century or two. And while fusion might be the energy source of the future, fission technology is already here.

Fission may not be the answer to humanity's energy needs; we might well have to turn to fusion in the more distant future. Nevertheless, from a purely practical point of view, fission seems to be a more reasonable solution than fusion, at least in the short term. Other, non-nuclear, possibilities exist as well. For example, if we figure out how to trap and sequester carbon dioxide, we might be able to burn coal and methane without releasing greenhouse gases. Carbon sequestration schemes and advanced fission reactor designs aren't sexy, cutting-edge science, but they are much more likely than fusion to help the next few generations of humans.

Even so, the fusion community clings to the hope that fusion energy is just thirty years away—and that it will solve *all* our energy problems. Despite the failures of the past, despite the enormous hurdles ahead, despite the tremendous cost, despite the easier alternatives, scientists still insist that fusion energy is the path forward. It is just another case of wishful thinking.

━━━━

There's something uniquely powerful about the promise of fusion energy. It harks back to the ancient quest to build a perpetual motion machine, but this time the source of unlimited energy doesn't violate the laws of physics. To anyone who could harness the energy of a miniature star, fusion promised power. Not only would it give the world endless electrical power, it would give power to its inventors. To some scientists, this meant financial power. Still others sought the power of fame. Some saw military and political power. The rewards are so great that they can blind the scientists on the quest.

This is not to say that fusion science is worthless. Far from it. Plasma physicists have figured out the inner workings of distant stars—how they live and die. It is no coincidence that some of the world's leading experts in stellar dynamics and supernova explosions are at Los Alamos, Livermore, and Princeton. Furthermore, fusion physicists are exploring new territory—they are looking at hotter, denser matter than anyone has yet examined—and scientists learn interesting things whenever they expand the boundaries of a field. Apparently, a pinch machine at Sandia Natural Laboratories has recently created a plasma hotter than a billion degrees. If true, it is a tremendous achievement that opens a new regime—much hotter even than the center of a star—to experimenters. That is an accomplishment in itself, but whenever physicists talk to the public about Sandia's Z machine, fusion energy madness seems to grip them. A 2007 Sandia press release promised that "fired repeatedly, the machine could well be the fusion machine that could form the basis of an electrical generating plant only two decades away."

It seems the wishful thinking is as strong as ever.

The promise of a fusion reactor a few decades away has been a cliché for a half century. Every time it is repeated, it just illuminates how generation after generation of scientists, drunk with the promise of personal glory and unlimited energy, keep forgetting the hard lessons learned by their predecessors. The quest to put a star in the bottle is intoxicating. Fusion might be the energy source of the future. If fusion scientists are unable to rid themselves of their intemperate self-deception, it always will be.

APPENDIX: TABLETOP FUSION

Thiago Olson was an ordinary teenage boy, for the most part. He had one oddity. There was something mysterious about what he did after school. The seventeen-year-old's friends had nicknamed Olson "the mad scientist." For good reason: in his basement, he was building his own fusion device.

On June 20, 2006, Olson told his fellow fusion enthusiasts about his homemade "fusor," cobbled together from parts taken from a defunct x-ray machine. "Yesterday I input power into my fusor for the first time," he wrote, adding that he was happy to see "the familiar purple plasma" glowing away through a viewing window. Over the next weeks, Olson steadily improved his six-foot-tall device, upgrading the system that handles the deuterium gas in the machine. Three months later, Olson was making national news. "Michigan Teen Creates Nuclear Fusion," the headlines blared.

Olson had, in fact, done it. A neutron counter implied that Olson's fusor was producing about 200,000 neutrons a second. And though a fusion device might seem like a scary thing to keep in one's basement, the fusor was perfectly safe. Once the headlines broke, two government radiation-safety officers and a fire marshal visited his home and gave the fusor a clean bill of safety.

On the surface, it seemed that Olson had succeeded where

Pons and Fleischmann had failed. He had come up with a cheap "tabletop" device that actually achieved nuclear fusion. The public reacted with astonishment, because the cold-fusion debacle seemed to prove that tabletop fusion was impossible. However, it is not that hard to build a tabletop fusion device; people have been doing it for decades.

Indeed, Utah's history with tabletop fusion goes back decades before the Pons and Fleischmann fiasco. The first person to achieve fusion with a cheap device was Utah born, a young man who grew up on a farm. His name was Philo Farnsworth.

Farnsworth is best known for inventing electronic television. As a young boy, he was plowing a potato field—back and forth, back and forth—when he was inspired with an idea. He could use the same back-and-forth motion to "dissect" a photographic image with a stream of electrons. Though it took years for him to perfect the device itself, at age fourteen Philo Farnsworth had invented a rudimentary television camera.

Farnsworth's device, known as the image dissector, first turned a picture into a set of electrons. Light causes a peculiar material—cesium oxide—to emit electrons, so an image shining on a plate of cesium oxide will change from a pattern of light and dark spots into a pattern of electrons streaming from the plate. Electrons, unlike photons, are strongly affected by electric and magnetic fields, and Farnsworth exploited this property by using electromagnetic fields to move the electron image back and forth, plowlike, over a detector. This allowed Farnsworth to convert an image into an electronic format that could then be transmitted over the airwaves. Though it was a relatively crude device, it worked. The age of electronic television had begun.* Unfortunately, a nasty patent battle ensued. Farnsworth won, but he never got rich

---

* There were earlier versions of television that dissected the image with mechanical spinning disks and other nonelectronic means. They were impractical, so Farnsworth's device was a major leap forward.

from his invention. (Just the opposite; it nearly drove him to madness. At one point, he committed himself to an insane asylum and received shock therapy.)

Farnsworth was a brilliant inventor, particularly when it came to manipulating charged particles—like electrons—with electric and magnetic fields. So when he first heard about attempts to create a fusion reactor in a magnetic bottle, he came up with the design for a device that he thought would do it. In the 1960s, he mortgaged his house and borrowed against his life insurance to make his dream a reality. The result was the Farnsworth fusor.

While Farnsworth's television camera manipulated electrons, his fusor manipulated deuterium nuclei. Stripped of its electrons, a deuterium nucleus has a charge equal and opposite to the electron's; though deuterium is much more massive than an electron, it, too, can be guided and accelerated by a powerful electric field.

Over the years, Farnsworth and his colleagues patented a number of slightly different designs for the fusor, but in principle they were all relatively simple. A fusor takes deuterium nuclei (or the nuclei of other elements) and injects them into a vacuum chamber that contains a pair of charged metal electrodes. The electrodes have to be shaped to allow the nuclei to pass through them; for example, they might be two concentric wire-mesh spheres, a big positively charged sphere surrounding a smaller negatively charged one. When a deuterium nucleus is squirted past the outer sphere, it is repelled by the positive charge and attracted to the negative charge of the inner sphere, so it zooms inward with ever increasing speed. If the spheres are kept at high voltages, the ions will be moving so fast that they will overshoot the inner sphere and plunge toward the center of the device, where they might strike other ions that have fallen inward from other directions. They might even fuse, releasing energy.

The fusor wasn't tough to build. The inventors had to be able to create a decent vacuum inside their chamber, construct electrodes designed to handle a very high voltage, and of course, get themselves some deuterium to inject into the device. Other than that, building the fusor was really pretty easy, thus well within the reach of a dedicated amateur. A small one can fit on a tabletop. And it works, too. Farnsworth got neutrons right away. Soon he and his colleagues were producing so many neutrons they had to run the device in a large pit, using the ground to shield them from the flood of particles.

Unfortunately, the Farnsworth fusor, as well as later devices that use electric fields to confine plasmas, will probably never be able to produce more energy than it gobbles up. Clever as the fusor design is, it is not a very good bottle for a star. Its electric fields let particles escape, and the motion of electrons in the plasma radiates energy away at an alarming rate. Nevertheless, fusors have acquired something of a cult following.

Young Thiago Olson's fusor—fundamentally the same as Farnsworth's device—is just one of more than a dozen that have sprung up in amateurs' basements around the country. Olson was the eighteenth amateur to achieve fusion on his own, according to a roll of honor on a Farnsworth fusor aficionado Web site that Olson regularly visited. (In fact, he wasn't the first high schooler on the list. The fifth amateur to achieve fusion, Tanhui Li, was also a high-school student; his fusor won him a scholarship in the 2003 Intel Science Talent Search.)* Though Olson doesn't make any claims that his device will solve the world's energy problems, many die-hard fusor fans are convinced, hoping against hope, that fusors will soon lead to a fusion reactor—a source of unlimited energy.

On November 9, 2006, just days before the Olson story broke, the fusion physicist Robert Bussard gave a talk at Google about his research with a modified fusor. He had been working for the navy, but after a number of years he had run out of money for the program. The scientist told his audience that if he could only get his hands on $200 million, he would be able to produce a working power plant within four to five years. Bussard was deceiving himself if mainstream scientific thought is any guide. The equations of plasma physics strongly imply that fusorlike devices are very unlikely ever to produce more energy than

---

* Li took second place in the competition, losing to another student who looked into treatments for yeast infections.

they consume. Nature's inexorable energy-draining powers are too hard to overcome.

Luckily, the fusor is not the only tabletop fusion device around. Plenty of researchers are building small, cheap fusion machines. Scientists without huge budgets have gotten fusion to work with inexpensive lasers, and by even stranger means.

A major hurdle with laser fusion is that electrons tend to absorb the light beam's energy better than the heavy nuclei they are attached to. But hot electrons are pretty much useless for inducing fusion, which requires hot, fast-moving nuclei instead. In an ingenious experiment, Todd Ditmire, a Livermore physicist, figured out how to turn this liability into an asset.

Ditmire injected microscopic droplets of deuterium into a vacuum chamber and then zapped them with a cheap infrared laser. Ordinary laser fusion scientists had long since abandoned infrared lasers because infrared light heats electrons too much. However, this effect was precisely what Ditmire was looking for. When he shot the laser at the deuterium microdroplets, the laser heated up their electrons, boiling them off in a fraction of a second. The positively charged nuclei left behind, stripped of their negatively charged electrons, began repelling their neighbors. All the nuclei immediately tried to escape from one another, and the droplets exploded with great force, spewing deuterium nuclei at high speeds in all directions. Ditmire's laser did the exact opposite of what traditional laser fusion was trying to do: instead of compressing and confining a dollop of deuterium plasma, he was causing it to blow apart. Ditmire discovered that on occasion, though, the fragments from exploding droplets—fast-moving deuterium nuclei—collide with each other and fuse. For every laser shot, he got about 1,200 neutrons from fusion. Considering that the energy of the laser was so low, less than what's put out by a Christmas light in a second, this was an impressive fusion yield. Even so, the energy produced by the fusion was ten million times less than the energy the laser poured in. Ditmire's scheme might be useful for studying fusion on a very tiny scale, but it will never lead to a reactor that produces more energy than it consumes.

Seth Putterman, working at UCLA, came up with an even more innovative way to induce fusion. He and his team created a device whose heart was made of a crystal of lithium tantalate, a compound with a very curious property. It is *pyroelectric*: when you heat it, electrons in the crystal rearrange themselves so that one side of the crystal is positively charged and the other side is negatively charged.* Attached to the crystal was a fine tungsten needle. When the researchers heated the lithium tantalate, the crystal's charges rearranged themselves and ran down the tungsten. The crystal and needle acted as a giant focusing device; all the energy of heating the crystal got turned into an extremely strong electric field right at the needle's tip.

Putterman and his colleagues put this device—about the size of a coffee can—in a chamber filled with deuterium. When they turned on the heater, the device worked as advertised, creating a huge electric field near the tip of the needle. When a deuterium atom ventured near the tip, the electric field immediately stripped off its electrons and sent the nucleus zooming away at tremendous speed toward a target filled with deuterium. On occasion, the flying deuterium would fuse with one in the target. All in all, the device produced about eight hundred neutrons per second. Again, it was an impressive display and might even lead to a commercial device to produce neutrons; however, it will always consume more energy than it produces. Beams of deuterium nuclei lose energy whenever they strike a target, and on average the amount of energy lost by the deuterium nuclei that don't fuse will outweigh the energy produced by those that do.

After the cold-fusion debacle, the idea of tabletop fusion seemed impossible—a pipe dream sought after only by cranks. (The first reaction of Michael Saltmarsh, the bubble fusion debunker, upon seeing Putterman's pyroelectric fusion paper was, "Oh, God, not again.") But in fact, tabletop fusion—fusion reactions carried out cheaply in a small

---

* This phenomenon is related to the more famous piezoelectricity. A piezoelectric crystal does the same thing—rearrange its charges—because of pressure changes rather than heat changes.

piece of laboratory equipment—is real, It just isn't yielding any more energy than it consumes, so it is useless as a source of power.

It is an unfortunate fact of nature: unless you are creating fusion in a hot, dense plasma, you are extraordinarily unlikely to produce excess energy. Too many phenomena conspire against you. Tabletop fusion is an interesting curiosity, but not a path to unlimited power.

ACKNOWLEDGMENTS

I have been covering fusion since I began my career as a science journalist, so it is impossible for me to thank all the people who have helped me understand the physics—and the politics—of the fusion quest. To all of them, even to those who might disagree with the conclusions of this book, I give my heartfelt gratitude.

I would also like to thank my editor, Wendy Wolf, as well as Hilary Redmon and Don Homolka for their help with the manuscript. I'm also grateful to my agents, John Brockman and Katinka Matson. My friends and colleagues in the journalism department at New York University have been wonderful to me, and I am grateful for their support. Finally, my friends and family have helped me tremendously. To my parents Burt and Tama, my brother Mark, and of course my wife, Meridith: thank you for everything.

NOTES

CHAPTER 1: THE SWORD OF MICHAEL

3 "The force from which the sun" Harold S. Truman, "White House Press Release on Hiroshima, August 6, 1945," in Cantelon, Hewlett, and Williams, eds., *The American Atom*, 65.

8n "pumpkin field" Argonne National Laboratory Web site, "The 'Last Universal Scientist' Takes Charge."

9 "I've done a terrible thing." Hijiya, "The *Gita* of J. Robert Oppenheimer," 150.

10 "He's a genius" As quoted in Rhodes, *The Making of the Atomic Bomb*, 448.

11 "He's a danger" Blumberg and Owens, *Energy and Conflict*, 1.

11 "The communists overturned every aspect" Teller, with Shoolery, *Memoirs*, 13, 15.

11n "In my acquaintance" *Time*, "Knowledge Is Power."

13n fellow physicists measured enthusiasm in "Tellers" Goodchild, *Edward Teller*, 127.

14 "I very soon found some unjustified assumptions in Teller's calculation" Rhodes, *The Making of the Atomic Bomb*, 419.

14n "On Edward Teller's blackboard at Los Alamos" Serber, *The Los Alamos Primer*, 4.

14n "I had become a bit annoyed with Fermi" Groves, *Now It Can Be Told*, 296-97.

15 "Edward first thought it was a cinch." Serber, *The Los Alamos Primer*, xxxi.

16 **Oppenheimer bet** Rhodes, *The Making of the Atomic Bomb*, 656, 668. Teller, with Shoolery, *Memoirs*, 211.

16 **"The time will come"** *Los Alamos Science*, "The Oppenheimer Years," 25.

16 **"There was no backing for the thermonuclear work"** Teller and Brown, *The Legacy of Hiroshima*, 41.

17 **"Edward offered to bet me"** John Manley quoted in Rhodes, *Dark Sun*, 404.

17 **"It is likely that a super-bomb can be constructed"** Ibid., 255.

18 **"a choice between the quick and the dead,"** Baruch, "The Baruch Plan."

18 **"We have evidence"** Truman, "Statement by President Truman, September 23, 1949."

19 **"Keep your shirt on!"** United States Atomic Energy Commission, *In the Matter of J. Robert Oppenheimer*, 714.

19 **"We are all agreed that it would be wrong"** General Advisory Committee to the U.S. Atomic Energy Commission, Report on the "Super," 30 October 1949, in Cantelon, Hewlett, and Williams, eds., *The American Atom*, 120.

19 **"A super bomb might become a weapon of genocide,"** Ibid., 121.

20 **"The fact that no limits exist to the destructiveness of this weapon"** Ibid., 122.

21 **"The day has been filled, too, with talk about supers,"** David E. Lilienthal, *The Journals of David E. Lilienthal*, Vol. 2, 577 (entry of 10 October 1949).

21 **"Now I began to see a distorted human being,"** McMillan, *The Ruin of J. Robert Oppenheimer*, 44–45.

21 **"[Scientists] are working and 'have made considerable progress'"** Friendly, "New A-Bomb Has 6 Times Power of 1st," 1.

22 **"It is part of my responsibility as Commander in Chief"** *Time*, "The Decision L Is Yes," 6 February 1950.

23 **"Every day Stan would come into the office,"** Rhodes, *Dark Sun*, 423.

23 **"pale with anger."** Herken, *Brotherhood of the Bomb*, 223.

23 **"took real pleasure"** Goodchild, *Edward Teller*, 163.

24 **"Teller was not easily reconciled to our results,"** Ulam, *Adventures of a Mathematician*, 216.

24 **"You can't get cylindrical containers of deuterium to burn"** Goodchild, *Edward Teller*, 166.

24 **"[Teller] was blamed at Los Alamos"** Bethe, "Comments on the History of the H-Bomb," 47.

25 **"The holiday is over,"** Reprinted in Shepley and Blair, *The Hydrogen Bomb*, 112.

25 "He proposed a number of complicated schemes" Bethe, "Comments on the History of the H-Bomb," 48.

25n After the Soviets and Chinese Cumings, *Origins of the Korean War*, 741.

26 "his report was focused on the Super and was so negative" Teller, with Shoolery, *Memoirs*, 303.

26 "in a very different tone" Teller, with Shoolery, *Memoirs*, 303.

26 "Edward is full of enthusiasm about these possibilities," Rhodes, *Dark Sun*, 467.

26n "Ulam felt that he invented the new approach to the hydrogen bomb." Ibid., 471.

27 The remaining 25 kilotons Ibid., 474.

27 "Eniwetok would not be large enough" Ibid., 424.

28 "I expected that the General Advisory Committee," United States Atomic Energy Commission, *In the Matter of J. Robert Oppenheimer*, 720.

28 "had created difficulties" Teller, with Shoolery, *Memoirs*, 327.

28n "technically sweet" United States Atomic Energy Commission, *In the Matter of J. Robert Oppenheimer*, 81.

29 "I am leaving the appeasers to join the fascists," Rhodes, *Dark Sun*, 496.

30 fourteen buildings the size of the Pentagon U.S. Department of Defense, Defense Nuclear Agency, *Operation Ivy*, p. 188.

30 When that dot of light Teller and Brown, *The Legacy of Hiroshima*, 56.

31 "It is my belief that if at the end of the war" Ibid., 714.

31 "go fishing for the rest of his life" Ibid., 721.

31 "In my new land," Teller, with Shoolery, *Memoirs*, 397.

31 "We believe that, had Dr. Oppenheimer given his enthusiastic support" United States Atomic Energy Commission, *In the Matter of J. Robert Oppenheimer*, 1017.

32 "disturbing." Ibid., 1019.

32 "interchangeable with the conventional weapons" Chang, "To the Nuclear Brink," 106.

32 "next month or two." Ibid., 107.

32n "Isn't it a fair statement today, Dr. Oppenheimer," United States Atomic Energy Commission, *In the Matter of J. Robert Oppenheimer*, 149.

CHAPTER 2: THE VALLEY OF IRON

37 "one of the gravest" Darwin, *The Origin of Species*, 505.

39n "If belief in the reality of atoms is so crucial," Blackmore, "Ernst Mach Leaves 'The Church of Physics,'" 524–25.

55 **As early as 1949, scientists realized** Reines and Suydam, *Preliminary Survey of Physical Effects Produced by a Super Bomb*, 910.

CHAPTER 3: PROJECT PLOWSHARE AND THE SUNSHINE UNITS

58 **"An underground explosion was indeed carried out"** U.S. Department of State, *Foreign Relations of the United States 1964–1968*, vol. 11, item 66.

59 **"the answer to a dream as old as man himself,"** *A is for Atom*, directed by Carl Urbano.

59 **"into the hands of those who will know how to strip its military casing"** Dwight D. Eisenhower, "The 'Atoms for Peace' Address to the United Nations General Assembly," 8 December 1953, in Cantelon, Hewlett, and Williams, eds., *The American Atom*, 102.

60 **"A Higher Intelligence decided"** Lewis Strauss, "My Faith in the Atomic Future," in Cantelon, Hewlett, and Williams, eds., *The American Atom*, 107.

60 **"If your mountain is not in the right place,"** O'Neill, *The Firecracker Boys*, 88.

60n **"I had a hand in formulating and popularizing that hope"** David E. Lilienthal, *Change, Hope, and the Bomb*, 109.

61 **"So you want to beat"** Teller and Brown, *The Legacy of Hiroshima*, 82.

62 **"they shall beat their swords into plowshares"** The King James Version of the Bible, Isaiah 2:4.

62 **The ideas started coming** Teller and Brown, *The Legacy of Hiroshima*, 80–91; O'Neill, *The Firecracker Boys*, 26; Seaborg, with Loeb, *Stemming the Tide*, 310; McPhee, *The Curve of Binding Energy*, 112–13.

62 **"We will change the earth's surface to suit us,"** Teller and Brown, *The Legacy of Hiroshima*, 84.

62 **"One will probably not resist for long"** O'Neill, *The Firecracker Boys*, 23–24.

63 **the harbor made little economic sense** Ibid., 39.

64 **"pinhead-sized white and gritty snow"** U.S. Department of Defense, Defense Nuclear Agency, *Castle Series, 1954*, 210.

64n **A Russian soldier died** Sakharov, *Memoirs*, 192.

65 **"raw, weeping lesions"** Cronkite et al., *Study of Response of Human Beings Accidentally Exposed to Significant Fallout Radiation*, 3–4.

65 **"well and happy"** *The Atomic Cafe*, directed by Jayne Loader and Kevin Rafferty.

65 **"Radioactive Fish Sought In Japan"** Parrott, "Nuclear Downpour Hit Ship During Test at Bikini—U.S. Inquiry Asked," 9.

65 "Each nuclear bomb test" Pauling, "Science and Peace."

66 "It is possible to say unequivocally" Johnston, "No Danger Seen in Nuclear Tests."

66 "observable fallout on Los Angeles." Ogle, *An Account of the Return to Nuclear Weapons Testing After the Moratorium 1958–1961*, 101.

66 "Radiation from test fallout" Teller and Brown, *The Legacy of Hiroshima*, 180.

66 "Our custom of dressing men in trousers" Ibid., 181.

66 Teller even suggested that the dead captain: Ibid., 173.

66n "The opposition Pauling encountered" Jahn, "Presentation Speech."

66n "So human samples are of prime importance," "In the Matter of: Biophysics Conference." Transcript, p. 8, lines 12–14.

67 "fallout fear-mongers": Teller and Brown, *The Legacy of Hiroshima*, 181.

67 "insignificant and doubtful medical considerations" Ibid., 183.

67 "The Administration was bracing itself today" Kenworthy, "U.S. Thinks Soviet Will Pledge Halt In Nuclear Tests," 1.

67 "cessation of tests of all forms" Associated Press, "Text of Resolution."

67 "Russia has beaten us on propaganda" Kenworthy, "U.S. Warns Free Nations Not to Be Misled by Soviet," 16.

67n "Deploring the mutations that may be caused by fallout" Teller and Brown, *The Legacy of Hiroshima*, 181.

68 "limited war" Ibid., 235ff.

69 "Beat your plowshares into swords," The King James Version of the Bible, Joel 3:10.

70 "grave consequences" Toth, "Teller Opposes Test Ban Treaty."

70 "you will have given away" Toth, "Teller Shows Consistency in Opposing Test Ban."

70 "superfluous and even dangerous" Seaborg, with Loeb, *Stemming the Tide*, 314.

70n "The Communists might develop Plowshare before we do," Teller and Brown, *The Legacy of Hiroshima*, 87.

72 "I've never seen [Teller] take a position" Blumberg and Owens, *Energy and Conflict*, 407.

72 "not acceptable to regional refineries." Nordyke, *The Soviet Program for Peaceful Uses of Nuclear Explosions*, 25.

72 the Soviets had to declare Ibid., 39.

CHAPTER 4: KINKS, INSTABILITIES, AND BALONEY BOMBS

74 "Among other bodies" Wilson, *Religio Chemici*, 264.

75 "The project is still in the early stages," Warren, "Perón Announces New Way to Make Atom Yield Power," 1.

76 "Foreign scientists will be interested to learn" Mariscotti, *The Secret of Huemul Island*, 8.

76 "a totally new way of obtaining atomic energy" Ibid., 9.

76 "several million degrees." Ibid., 8.

76 "What the Americans get" Ibid., 9.

76 "Yes, sir, for the very first time" Ibid., 180.

77 "Less than that." *New York Times*, "Lilienthal Scouts Claim."

77 "It is strange that the names of eminent physicists" Coseia Ribeiro quoted in Associated Press, "Discoverer of Argentina's Atomic Energy Method Gives Few Details."

77 "I know what that other material is" *Time*, "Perón's Atom."

77 "Baloney bomb" *Time*, "Energy of the Pampas."

77 "I am not interested" Warren, "Perón Is Scornful of Atomic Skeptics," 1.

77 "They have not yet told the first truth," Ibid.

77n "There seemed to be some puzzlement," Ibid., 11.

78 "so-so" *Time*, "Energy of the Pampas."

79 "striking similarity" *New York Times*, "French Atom Expert Backs Perón Claim."

79 "Richter admits that his process is not new," Kaempffert, "Argentina Lacks Development Resources Even If, in Theory, Its Atomic Tests Are Possible."

79 "said that Dr. Ronald Richter, the former Austrian scientist," *Time*, "On Further Examination . . ."

80 "(a) Perón has fallen victim to a crank" Thirring, "Is Perón's A-Bomb a Swindle?" 2.

80 "The reactor operation crew and I are deeply sorry" Richter, "Argentina Has No Atom Bomb," i.

80 "It must really have been the deepest degree of secrecy" Ibid., iii.

80 "in full-scale operation" Ibid., ii.

80 "would be able to make convincing new demonstrations" Associated Press, "Argentine Plans Atomic Trade."

81 He was shocked when Richter Mariscotti, *The Secret of Huemul Island*, 280ff.

81 The visit was a fiasco. Ibid., 287ff.

88 "I venture to predict" Miyahara, "A Record of the First Five Meetings of 'The Symposium on Fusion Technology Held between 1960 and 1968,'" 384.

91 "They try to snap inward" Teller, *Energy From Heaven and Earth*, 209.

96n pale ale: Carruthers, "The Beginning of Fusion at Harwell," 1998.

97 "The indications are that fusion has been achieved" United Press, "Britons Report Gain on Fusion Reaction."

98n "backlog of experimental results" *New York Times*, "British Deny U.S. Gags Atomic Gain."

99 "Admiral Strauss' tactics" Love, "British-U.S. Data on Hydrogen Due."

99 "90% certain" Love, "Briton 90% Sure Fusion Occurred."

99 "Gains in Harnessing Power" Finney, "Gains in Harnessing Power of H-Bomb Reported Jointly by U.S. and Britain," 1.

99 "UNLIMITED POWER from SEA WATER" Carruthers, "The Beginning of Fusion at Harwell," 1999.

99 "achievement in harnessing thermonuclear energy" *New York Times*, "Soviet Radio Hails British on Fusion."

99n "British scientists pointed out" United Press, "Moscow Takes a Bow."

100 "Britain Indicates Reactor Advance" Love, "Britain Indicates Reactor Advance."

100 "H-Bomb Untamed, Britain Admits." Love, "H-Bomb Untamed, Britain Admits."

101 "It is doing exactly the job we expected" Associated Press, "Fusion Tests Upheld."

CHAPTER 5: HEAT AND LIGHT

102 "What glory beats" Buch, *Richter: Ópera Documental de Cámara.*

103 "looks probable as a thermonuclear source." Finney, "U.S. Experts Hint H-Bomb Control.

103 "We are now prepared to stake our reputations" *Time*, "On the Way: Genuine Fusion."

106 "partly as a research facility," Bromberg, *Fusion*, 64.

107 "entirely as a research facility," Ibid.

108 "I was so stunned" Sakharov, *Memoirs*, 92.

108n "One had only to substitute 'USSR' for 'USA,'" Ibid., 100.

111 One disruption Seife, "At the Going Down of the Nuclear Sun," 95.

113n "I expect that they use a very long thermometer." Braams and Stott, *Nuclear Fusion*, 135.

116 One Atomic Energy Commission worker: Bromberg, *Fusion*, 167.

118 "a significant step toward the long-range goal" Schmeck, "Nuclear Fusion Reported In Lab With Aid of Laser."

118 "a small but significant initial step" Ibid.

118 "efficient fusion power" Bromberg, *Fusion*, 186.

CHAPTER 6: THE COLD SHOULDER

127 "We are also human, and we need miracles," Garwin, "Consensus on Cold Fusion Still Elusive," 617.

128 "We used to call it" Thomas Valone, Second International Conference on Future Energy, 24 September 2006.

128 "Two scientists have successfully created" Huizenga, *Cold Fusion*, 237.

129 "ranks right up there with fire," Taubes, *Bad Science*, xx.

131 They mused about it Bishop and Wells, "Taming H-Bombs?"

131 "Stan and I thought this experiment was so stupid" Fleischmann, Pons, and others, press conference of 23 March 1989.

132 "Sometime during the night" Bishop, "Heat Source in Fusion Find May Be Mystery Reaction," 1.

133 "interesting discussion" Alvarez et al., "Catalysis of Nuclear Reactions by μ Mesons," 1128.

134 "in effect, confines the two nuclei in a small box." Ibid.

135 "We had a short but exhilarating experience" Alvarez, "Recent Developments in Particle Physics."

136 "each muon may catalyze hundreds" Jones, "Muon-Catalyzed Fusion Revisited," 132.

136 "It is now conceivable" Rafelski and Jones, "Cold Nuclear Fusion," 84.

137 "a bunch of baloney," Taubes, *Bad Science*, 25.

138 the competitors had agreed Footlick, *Truth and Consequences*, 35.

138 Pons and Fleischmann jumped the gun: Taubes, *Bad Science*, 87ff.

138 "Stan and I often talk" Huizinga, *Cold Fusion*, 238.

141 "provides strong evidence" Jones et al., "Observation of Cold Nuclear Fusion in Condensed Matter," 739.

144 "It's wrong." Taubes, *Bad Science*, 141.

145 "There is big money in hot fusion," Footlick, *Truth and Consequences*, 49.

145 "The black hats, such as they were," Ibid., 44.

145 "Within the next few weeks," Garwin, "Consensus of Cold Fusion Still Elusive," 617.

146 "Now it appears that chemists" Pool, "Skepticism Grows Over Cold Fusion," 284.

146 "We do not get the total blank experiment" Ibid. 285.

147 "How is this astounding oversight" Maddox, "What to Say about Cold Fusion," 701.

147 "urgent work." *Nature*, "Cold Fusion in Print," 604.

147 "in a way that maximized publicity" Bazell, "Hype-Energy Physics," 7.

147 "The pace of scientific advance" *Wall Street Journal*, "Fusing the Impossible," 1.

149 "sure as sure can be" Lindley, "More Than Scepticism," 4.

149 "By the time the hearings got around to the skeptics," Park, "A Cold Fusion Institute Was Called for in Hearings."

150 "polite but generally sceptical reception," Lindley, "More Than Scepticism," 4.

150 "We're suffering from the incompetence and delusions" Waldrop, "Cold Water from Caltech," 523.

150 "We asked Pons if he stirred," Ibid.

150 "devastating" Bishop, "Physicists Outline Possible Errors That Led to Claims of Cold Fusion," 1.

151 "continue to twitch for a while." Park, "The Corpse of Cold Fusion Will Probably Continue to Twitch."

151 John Sununu abruptly cancelled Taubes, *Bad Science*, 267

151 Pons and Fleischmann got more reclusive Ibid., 293, 375ff, 421.

151 Pons's lawyer threatened legal action Ibid., 355.

151 "I discovered after awhile" Footlick, *Truth and Consequences*, 42.

152 "modest support" U.S. Department of Energy, Energy Research Advisory Board, *Cold Fusion Research*.

154 he tried to prove Jones, "Behold My Hands: Evidence for Christ's Visit in Ancient America"; Jones, "Why Indeed Did the WTC Buildings Completely Collapse?"

154 "These folks need a fair hearing," Siegel, "America Condemns Inventive Science, Fathers of Cold Fusion Tell ABC," B1.

154 "a colossal conspiracy of denial." Platt, "What If Cold Fusion Is Real?"

154 "Cold fusion today is a prime example of pathological science," Begley, "Cold Fusion Isn't Dead, It's Just Withering from Scientific Neglect," B1.

155 "Valone called for" Voss, "A Free Energy Enthusiast Seeks Like-Minded Colleagues," 1254.

155 "The speakers list for CoFE" Park, "Free Energy: State Department Opens Its Doors to New Agers."

156 "In cooperation with the U.S. Department of Commerce" Park, "Future Energy: New Age Energy Conference Moves to Commerce."

156 "The federal government's budget pie" "In the Matter of Arbitration Between Patent Office Professional Association and U.S. Department of Commerce, Patent and Trademark Office."

CHAPTER 7: SECRETS

160 "the breaking of an enemy code" Bylinsky, "Shiva: The Next Step to Fusion Power," 87.

160 "energy release at the giant $25 million Shiva" Sullivan, "Reports Assay Failures and Hopes for Fusion Power."

162 80 percent of the capsules failed to ignite, Seife, "Will NIF Live Up to Its Name," 1128.

164 "the operation of a magnetic fusion demonstration plant" *Magnetic Fusion Energy Engineering Act of 1980*, §2 (b) 4.

165n "I thought I was going to a wake" Herman, *Fusion*, 209.

167 Rush Holt Presentation to college group, 5 October 1992.

CHAPTER 8: BUBBLE TROUBLE

170 "Hegel observes somewhere" Marx, *The 18th Brumaire of Louis Napoleon*, 287.

172 "Here's a *Science* paper" Robert Coontz, e-mail to author, 5 February 2002.

176 "Right after you did your story" Rusi Taleyarkhan, e-mail to author, 6 February 2002.

180 "I thought, doggone!" Seife, "'Bubble Fusion' Paper Generates Tempest in a Beaker," 1808.

181 "There was certainly pressure from Oak Ridge" Ibid., 1809.

182 "I understand there has been some discussion" Richard Garwin, e-mail message, 28 February 2002.

182 "I am told that the paper in question," Will Happer, e-mail message, 1 March 2002.

183 "I like *Science*; I'm a member of AAAS," Will Happer, telephone conversation with author, 2002.

183 "It would be unfortunate" Richard Garwin, telephone conversation with author, 2002.

183   "Concern on the part of research managers" Don Kennedy, e-mail message, (forwarded by Happer), 26 February 2002.

184   "I'm confused" Draft of Seife, "'Bubble Fusion' Paper Generates Tempest in a Beaker," 4 March 2002.

184   "The paper's kind of a patchwork" Ibid.

185   "calling around asking about activities" E-mail extract (unidentified) forwarded by Will Happer to author, 1 March 2002.

186   "Be careful what you write" Will Happer, e-mail to author, 1 March 2002.

186   "Taleyarkhan and his team," Draft of Seife, "'Bubble Fusion' Generates Tempest in a Beaker," 4 March 2002.

187   "Bubble Fusion: A Collective Groan" Park, "Bubble Fusion: A Collective Groan Can Be Heard."

187   "Fusion in a Flash?" American Association for the Advancement of Science, "Fusion in a Flash?"

189   "Preliminary Evidence Suggests" Oak Ridge National Laboratory, "Preliminary Evidence Suggests Possible Nuclear Emissions during Experiments."

190   "I'm sorry if my cold-fusion allergy" Don Kennedy, e-mail to author, 4 March 2002.

190   "[W]e ask all Science Press Package registrants" American Association for the Advancement of Science, "Special Notice to All Science Press Package Recipients."

191   "Fusion 'Breakthrough' Heralds Cleaner Energy," Connor, "Fusion 'Breakthrough' Heralds Cleaner Energy," 9.

191   "Here we go again; Table-top fusion" Economist, "Here We Go Again; Table-Top Fusion," 95.

192   "I reviewed the paper twice," Vedantam, "Fusion Experiment Sparks an Academic Brawl," A10.

192   "unready for publication" Putterman, Crum, and Suslick, "Comments on 'Evidence for Nuclear Emissions During Acoustic Cavitation'; by R. P. Taleyarkhan et al., 1.

193   "Somewhere out there is a positive report" Adam and Knight, "Publish, and Be Damned . . . ," 774.

193   "We maintain our end" Ibid.

194   "They're saying, 'We understand'" Seife, "Chemistry Casts Doubt on Bubble Reactions."

194   "Evidence Bubbles Over" Purdue University, "Evidence Bubbles Over to Support Tabletop Nuclear Fusion Device."

195 "Purdue Findings Support" Purdue University, "Purdue Findings Support Earlier Nuclear Fusion Experiments."

195 "independently confirmed." Taleyarkan et al., "Nuclear Emissions During Self-Nucleated Acoustic Cavitation," 034301-1.

196 "Nuclear fusion is a major finding" "An Experiment to Save the World," narrated by Dilly Barlow.

197 "I would help out anybody" Ibid.

197 "He said: 'Look, there's a peak,'" Reich, "Is Bubble Fusion Simply Hot Air?"

198 Taleyarkhan is said to have pressed Ibid.

198 "nothing but a contrived and hurried attempt" Dunn, "RE: Inquiry Committee Proceedings Under Executive Memorandum No. C-22."

198 "Presenting that as independent is fraud" Maugh, "College Reviews Physicist's Tabletop Fusion Claims," A9.

198 "negligent or jumped the gun" Ibid.

198 "data from Xu's paper" Reich, "Purdue Attacked over Fusion Inquiry," 664.

199 "a severe lack of judgment" Dunn, "RE: Inquiry Committee Proceedings Under Executive Memorandum No. C-22."

199 "Despite the University's statement" U.S. Congress, House Committee on Science and Technology, Subcommittee on Investigations and Oversight, letter to Martin C. Jischke, 27 May 2007, 1.

199 "several matters warrant further investigation" Purdue University, "Panel Determines Research Allegations Merit Additional Investigation."

199 "I am exhausted," Rusi Taleyarkhan, telephone conversation with author, 29 August 2007.

200 Taleyarkhan had committed scientific misconduct Purdue University. "Report of the Investigation Committee In the Matter of Rusi P. Taleyarkhan." Final Report of the C-22 Investigation Committee, 18 April 2008.

200 Taleyarkhan had deceived the scientific community Ibid.

200 deemed a deliberate attempt Ibid.

200 "was apparently done by one person" Ibid.

CHAPTER 9: NOTHING LIKE THE SUN

204 "We find that this cannot be done." President's Committee of Advisors on Science and Technology (PCAST), *The U.S. Program of Fusion Energy Research and Development*, 4.

206 "after ten years" Fusion Power Associates, "U.S. Fails to Sign ITER Extension Agreement; JET Demonstrates Fusion Remote Maintenance."

207 "breakeven plasma conditions" Japan Atomic Energy Research Institute, *Persistent Quest—Research Activities, 1997.*

207 "Breakeven was demonstrated" Parliamentary Office of Science and Technology, "Nuclear Fusion."

208 "Once we crack the problem of fusion," Angier, "High Hopes for a Super Nova."

212 "Not 1%. Those who say 5%" Seife, "Will NIF Live Up to Its Name?" 1128.

213 "but its contribution is almost exclusively to the long-term tasks" JASON, *Inertial Confinement Fusion (ICF) Review,* 2.

214 "NIF will contribute to training and retaining expertise" JASON, *Science Based Stockpile Stewardship,* 54.

217n "abundant resources" Bush, Speech Announcing New Vision for Space Exploration Program.

218 "hand held out to future generations" Chirac, Speech by Mr Jacques Chirac, President of the French Republic.

219 "The ambition is huge!" Ibid.

CHAPTER 10: THE SCIENCE OF WISHFUL THINKING

220 "When one turns to the magnificent edifice" James, *The Will to Believe,* 7.

222 "It was the thing that one fears" Seife, "CERN's Gamble Shows Perils, Rewards of Playing the Odds," 2261.

222 "We did all sorts of tests on the data" Ibid.

223 "Ten or twelve days before I was due" Ibid.

223 "It was a large audience" Ibid.

223 "Here I was, with the biggest blunder of my life" Ibid.

227 "fired repeatedly, the machine could well be the fusion machine" Sandia National Laboratories, "Rapid-Fire Pulse Brings Sandia Z Method Closer to Goal of High-Yield Fusion Reactor."

APPENDIX: TABLETOP FUSION

229 "the mad scientist" Damron, "Oakland Twp. Teen Creates Fusion."

229 "Yesterday I input power" Olson, "Archived—First Plasma."

229 "Michigan Teen Creates Nuclear Fusion" United Press International, "Michigan Teen Creates Nuclear Fusion."

234 "Oh, God, not again." Seife, "Tabletop Accelerator Breaks 'Cold Fusion' Jinx But Won't Yield Energy, Physicists Say," 613.

BIBLIOGRAPHY

A is for Atom. Directed by Carl Urbano. John Sutherland Productions, Inc., 1952.

Adam, David, and Jonathan Knight. "Publish, and Be Damned..." *Nature* 419 (24 October 2002): 772–76.

"Agreement on the Establishment of the ITER International Fusion Energy Organization for the Joint Implementation of the ITER Project." International agreement, 21 November 2006. http://www.iter.org/pdfs/Agreement.pdf.

Allardice, Corbin, and Edward R. Trapnell. "The First Pile." From U.S. Department of Energy, *The First Reactor*. DOE/NE-0046. Washington, DC: U.S. Department of Energy, 1982.

Allen, N. L., T. E. Allibone, D. R. Chick, et al. "A Stabilized High-Current Toroidal Discharge Producing High Temperatures." *Nature* 181 (25 January 1958): 222–24.

Allibone, T. E. "Fusion of Heavy Hydrogen." *Nature* 183 (28 February 1958): 569–73.

Alvarez, Luis W. "Recent Developments in Particle Physics." Nobel Prize Lecture, 11 December 1968. From *Nobel Lectures, Physics 1963–1970*. Amsterdam: Elsevier, 1972. http://nobelprize.org/nobel_prizes/physics/laureates/1968/alvarez-lecture.pdf.

Alvarez, L. W., H. Bradner, F. S. Crawford, et al. "Catalysis of Nuclear Reactions by μ Mesons." *Physical Review* 105 (February 1957): 1127–28.

American Association for the Advancement of Science. "Fusion in a Flash?" Press release. 4 March 2002.

———. "Special Embargo Notice." E-mail to journalists. 4 March 2002.

———. "Special Notice to All Science Press Package Recipients." E-mail to journalists. 4 March 2002.

"An Experiment to Save the World." Narrated by Dilly Barlow. BBC/Horizon, 2005. Transcript. http://www.bbc.co.uk/sn/tvradio/programmes/horizon/experiment_trans.shtml.

Anastasio, Mike. "Fusion Ignition as an Integrated Test of Stockpile Stewardship." *Science & Technology Review*, July–August 1999, 3.

Angier, Natalie. "High Hopes for a Super Nova." *Time*, 15 April 1985.

Argonne National Laboratory Web site. "The 'Last Universal Scientist' Takes Charge." http://www.anl.gov/Science_and_Technology/History/Anniversary_Frontiers/unisci.html (accessed 27 July 2007).

Associated Press. "Argentine Atomic Story Jolts American Foreign Ministers." *Lowell Sunday Sun*, 25 March 1952.

———. "Argentine Expert Answers Critics of Atomic Claims." *Winnipeg Free Press*, 7 July 1951.

———. "Argentine Hints at Atomic Trade." *New York Times*, 26 June 1951.

———. "Argentine Plans Atomic Trade." *New York Times*, 12 December 1951.

———. "Discoverer of Argentina's Atomic Energy Method Gives Few Details." *Newport (Rhode Island) Daily News*, 26 March 1951.

———. "Economical Power From Atom Is Not Envisioned Before 1965." *New York Times*, 17 October 1959.

———. "Foreign Reporters Excluded." *New York Times*, 25 March 1951.

———. "Fusion Tests Upheld." *New York Times*, 16 June 1958.

———. "More Information On What Argentina Uses for A-Power." *Kingsport (Tennessee) News*, 28 March 1951.

———. "Navy, in Fusion Tests, Heats Gas to 1,000,000°." *New York Times*, 29 March 1958.

———. "Perón Claims New Atomic Development." *Zanesville (Ohio) Signal*, 24 March 1951.

———. "Perón 'Not Interested' in What U.S. Thinks of Atom Discovery." *Austin (Minnesota) Daily Herald*, 26 March 1951.

———. "Peronist Scientist Freed." *New York Times*, 7 October 1955.

———. "Perón's Aide Denies Dropping Atom Plan." *New York Times*, 6 December 1952.

———. "Savant to Honor Peróns." *New York Times*, 25 May 1951.

———. "Text of Resolution." *New York Times*, 1 April 1958.

The Atomic Cafe. Directed by Jayne Loader and Kevin Rafferty. The Archives Project, 1982.

"Atomic Energy." *Nature,* 5 April 1958, 948.

"The Atomic Energy Society." *Nature,* 22 August 1959, 602–3.

Back, C. A., J. F. Davis, C. D. Decker, et al. *High-Power Laser Source Evaluation.* UCRL-ID-129096-98. Livermore, CA: Lawrence Livermore National Laboratory, 1998.

Bailes, M., A. G. Lyne, and S. L. Shemar. "A Planet Orbiting the Neutron Star PSR1829-10." *Nature* 352 (25 July 1991): 311–13.

Bainbridge, K. T. *Trinity.* LA-6300-H. Los Alamos, NM: Los Alamos Scientific Laboratory, 1976.

Barnard, Chester I., J. R. Oppenheimer, Charles A. Thomas, Harry A. Winne, and David E. Lilienthal. *A Report on the International Control of Atomic Energy* (Acheson-Lilienthal Report). Department of State Pub. 2498. Washington, DC: U.S. Government Printing Office, 1946. Available at http://www.learn-world.com/ZNW/LWText.Acheson-Lilienthal.html.

Baruch, Bernard. "The Baruch Plan." Speech to the United Nations Atomic Energy Commission, 14 June 1946. Available at http://www.streitcouncil.org/content/pdf_and_doc/The%20Baruch%20Plan.pdf.

Bazell, Robert. "Hype-Energy Physics." *New Republic,* 24 April 1989, 7–8.

Begley, Sharon. "Cold Fusion Isn't Dead, It's Just Withering from Scientific Neglect." *Wall Street Journal,* 5 September 2003.

Bernstein, Jeremy. *Oppenheimer: Portrait of an Enigma.* Chicago: Ivan R. Dee, 2004.

Bethe, Hans. "Comments on the History of the H-Bomb." *Los Alamos Science,* Fall 1982, 42–53.

———. "Energy Production in Stars." Nobel Prize Lecture, 11 December 1967. From *Nobel Lectures, Physics 1963–1970.* Amsterdam: Elsevier, 1972. http://nobelprize.org/nobel_prizes/physics/laureates/1967/bethe-lecture.pdf.

Bethe, Hans, K. Fuchs, J. O. Hirschfelder, J. L. Magee, R. E. Peieris, J. von Neumann. *Blast Wave.* LA-2000. Los Alamos, NM: Los Alamos Scientific Laboratory, 1947.

Bickerton, R. "Thermonuclear Processes." *Nature* 184 (25 July 1959): 240–41.

Bird, Kai, and Martin J. Sherwin. *American Prometheus: The Triumph and Tragedy of J. Robert Oppenheimer.* New York: Knopf, 2005.

Bishop, Amasa S. *Project Sherwood: The U. S. Program in Controlled Fusion.* Reading, MA: Addison-Wesley, 1958.

Bishop, Jerry E. "Heat Source in Fusion Find May Be Mystery Reaction." *Wall Street Journal,* 3 April 1989.

———. "Physicists Outline Possible Errors That Led to Claims of Cold Fusion." *Wall Street Journal,* 3 May 1989.

——. "Scientist Sticks to Test-Tube Fusion Claim." *Wall Street Journal*, 27 March 1989.

Bishop, Jerry E., and Ken Wells. "Taming H-Bombs?" *Wall Street Journal*, 24 March 1989.

Blackmore, John. "Ernst Mach Leaves 'The Church of Physics.'" *British Journal of the Philosophy of Science* 40 (1989): 519–40.

Blakeslee, Alton. "Sea Atoms May Solve Mankind's Power Problems." Associated Press. *Benton Harbor (Michigan) News-Palladium*, 28 August 1958.

Blumberg, Stanley A., and Gwinn Owens. *Energy and Conflict: The Life and Times of Edward Teller.* New York: G. P. Putnam's Sons, 1976.

Braams, C. M., and P. E. Stott. *Nuclear Fusion: Half a Century of Magnetic Confinement Fusion Research.* Bristol: Institute of Physics Publishing, 2002.

Broad, William. "Cold Fusion Claim Is Faulted on Ethics as Well as Science." *New York Times*, 17 March 1991.

Bromberg, Joan Lisa. *Fusion: Science, Politics, and the Invention of a New Energy Source.* Cambridge, MA: MIT Press, 1982.

Brown, P. S., and L. J. Ferderber. *The Stockpile Stewardship Program.* UCRL-JC-131080. Livermore, CA: Lawrence Livermore National Laboratory, 1993.

Browne, Malcolm W. "Fusion Power: Is There Still an El Dorado for Energy?" *New York Times*, 15 April 1979.

——. "Physicists Debunk Claim of a New Kind of Fusion." *New York Times*, 3 May 1989.

Brumfiel, Geoff. "Bubble Fusion Dispute Reaches Boiling Point." *Nature* 416 (7 March 2002): 7.

——. "Lack of Funding Puts 'Bubble Fusion' Replication on Hold." *Nature* 417 (27 June 2002): 887.

——. "Misconduct? It's All Academic…" *Nature* 445 (18 January 2007): 240–41.

Buch, Esteban. *Richter: Ópera Documental de Cámara.* Libretto (Unpublished copy supplied by librettist; translated by Meridith Seife and Charles Seife).

Burchfield, Joe D. "Darwin and the Dilemma of Geological Time." *Isis* 65 (September 1974): 300–321.

Burkhardt, L. C., and R. H. Lovberg. "New Confinement Phenomena and Neutron Production in a Linear Stabilized Pinch." *Nature* 181 (25 January 1958): 228–30.

Burkhardt, L. C., R. H. Lovberg, and J. A Phillips. "Magnetic Measurement of Plasma Confinement in a Partially Stabilized Linear Pinch." *Nature* 181 (25 January 1958): 224–25.

Bush, George W. Speech Announcing New Vision for Space Exploration Program. NASA Headquarters, Washington, DC, 14 January 2002. http://whitehouse.gov/news/releases/2004/01/20040114-3.html.

Bussard, Robert. "Should Google Go Nuclear?" Google Tech Talk, 9 November 2006. http://video.google.com/videoplay?docid=1996321846673788606 (accessed 2 July 2007).

Bylinsky, Gene. "Shiva: The Next Step to Fusion Power." *Fortune,* 30 January 1978, 85–88.

Cantelon, Philip L., Richard G. Hewlett, and Robert C. Williams, eds. *The American Atom.* 2nd ed. Philadelphia: University of Pennsylvania Press, 1991.

Carey, Frank. "Argentina Claims New Atom Secret." Associated Press. *Council Bluffs (Iowa) Nonpareil,* 26 March 1951.

Carruthers, R. "The Beginning of Fusion at Harwell." *Plasma Physics and Controlled Fusion* 30, no. 14 (1998): 1993–2001.

Cartlidge, Edwin. "Europe Plans Laser-Fusion Facility." *PhysicsWeb,* 2 September 2005. http://physicsweb.org/articles/news/9/9/2/1 (accessed 25 July 2007).

Cartwright, David C. *Inertial Confinement Fusion at Los Alamos.* Vol. 1, *Progress in Inertial Confinement Since 1985.* LA-UR-89-2675. Los Alamos, NM: Los Alamos National Laboratory, 1989.

Chadwick, J. "The Existence of a Neutron." Originally in *Proceedings of the Royal Society A* 136 (1932): 692–708. Available at http://dbhs.wvusd.k12.ca.us/webdocs/Chem-History/Chadwick-1932/Chadwick-neutron.html.

Chang, Gordon. "To the Nuclear Brink: Eisenhower, Dulles, and the Quemoy-Matsu Crisis." *International Security* 12 (Spring 1998): 96–123.

Chang, Kenneth. "Experts Say New Desktop Fusion Claims Seem More Credible." *New York Times,* 3 March 2004.

Chirac, Jacques. Speech by Mr Jacques Chirac, President of the French Republic, On the occasion of the signing ceremony for the International ITER Agreement on controlled Nuclear Fusion, Paris, Elysée Palace, 21 November 2006. http://www.iter.org/pdfs/speech_Chirac.pdf.

Clery, Daniel. "ITER Pact Signed." *ScienceNOW,* 21 November 2006. http://sciencenow.sciencemag.org/cgi/content/full/2006/1121/1 (accessed 22 November 2006).

Close, Frank. *Too Hot to Handle: The Race for Cold Fusion.* Princeton: Princeton University Press, 1991.

Cockcroft, J. D. "Peaceful Uses of Atomic Energy: United Nations Conference at Geneva." *Nature* 182 (4 October 1958): 903–5.

Cohen, James S, and John D. Davies. *"The Cold Fusion Family."* Nature 338 (27 April 1989): 705–7.

Coleman, L. *Status of Inertial Fusion in the United States.* UCRL-JC-108463. Livermore, CA: Lawrence Livermore National Laboratory, 1991.

Connor, Steve. "Fusion 'Breakthrough' Heralds Cleaner Energy." *Guardian,* 6 March 2002.

Cowan, George A. *Nuclear Explosions as Neutron Sources.* LAMS-2391. Los Alamos, NM: Los Alamos Scientific Laboratory, 1959.

Crawford, Mark. "Cold Fusion: Is It Hot Enough to Make Power?" *Science* 244 (28 April 1989): 423.

——. "Utah Looks to Congress for Cold Fusion Cash." *Science* 244 (5 May 1989) 522–23.

Cronkite, E. P., V. P. Bond, L. E. Browning, et al. *Study of Response of Human Beings Accidentally Exposed to Significant Fallout Radiation.* Operation Castle—Final Report Project 4.1. Naval Medical Research Institute, Bethesda, MD, and U. S. Naval Radiological Defense Laboratory, San Francisco, CA, October 1954.

Crum, Lawrence A. "Sonofusion: Star in a Jar?" PowerPoint Presentation. http://www.washington.edu/research/graphics/sf/pdfs/Crum.pdf (accessed 26 July 2007).

Cumings, Bruce. *The Origins of the Korean War.* Vol. 2, *The Roaring of the Cataract, 1947–1950.* Princeton: Princeton University Press, 1990.

Curie, (Marie) Sklowdowska. "Radium and Radioactivity." Originally in *Century Magazine,* January 1904, 461–66. http://www.aip.org/history/curie/article_text.htm (accessed 25 July 2007).

Damron, Gina. "Oakland Twp. Teen Creates Fusion." *freep.com,* 19 November 2006. http://www.freep.com/apps/pbcs.dll/article?AID=2006611190602 (accessed 21 November 2006; currently defunct).

Darwin, Charles Robert. *The Origin of Species.* The Harvard Classics. New York: P. F. Collier & Son, 1909.

Davies, N. Anne. Letter to John A. Woods, 2 October 1998 (copy supplied to author).

The Day After Trinity. Directed by John Else. KTEH, 1981.

de Hoffmann, Frederic. *Minutes of an Informal Meeting on Nuclear Rockets.* LAMS-836. Los Alamos, NM: Los Alamos Scientific Laboratory, 1949.

Dean, Steven O. "Fifty Years of U.S. Fusion Research—An Overview of Programs." *Nuclear News,* July 2002, 34–40.

——. "Status and Objectives of Tokamak Systems for Fusion Research." *Journal of Fusion Energy* 17, no. 4 (1998): 289–337.

——. "The Decision L Is Yes." 6 February 1950.

Dibble, Timothy, Saibal Bandyopadhyay, Jamal Ghoroghchian, et al. "Electrochemistry at Very High Potentials: Oxidation of the Rare Gases and Other Gases in Nonaqueous Solvents at Ultramicroelectrodes." *Journal of Physical Chemistry* 90 (9 October 1986): 5275–77.

Dickman, Steven. "1920s Discovery, Retraction." *Nature* 338 (27 April 1989) 692.

Ditmire, T., J. Zweiback, V. P. Yanovsky, T. E. Cowan, G. Hays, and K. B. Wharton. "Nuclear Fusion from Explosions of Femtosecond Laser-Heated Deuterium Clusters." *Nature* 398 (8 April 1999) 489–92.

Divine, Robert A. *Blowing on the Wind: The Nuclear Test Ban Debate*. New York: Oxford University Press, 1978.

——. *Eisenhower and the Bomb*. New York: Oxford University Press, 1981.

Duncan, Mark. "Should Google Go Nuclear?" 7 May 2007. http://www.askmar. com/ConferenceNotes/Should%20Google%20Go%20Nuclear.pdf (accessed 24 July 2007).

Dunn, Peter H. "RE: Inquiry Committee Proceedings Under Executive Memorandum No. C-22." Letter to Rusi Taleyarkhan and Leah H. Jamieson, Dec. 15, 2006.

Economist. "Here We Go Again; Table-Top Fusion." 9 March 2002, 95.

Eder, G., and H. Motz. "Contribution of High-Energy Particles to Thermonuclear Reaction Rates." *Nature* 182 (28 October 1958): 1140–42.

Einstein, Albert. Letter to F. D. Roosevelt, 2 August 1939. Available at http:// www.anl.gov/Science_and_Technology/History/Anniversary_Frontiers/ aetofdr.html.

Emmett, John L., et al. "NIF Technology Review." Report of the Technology Resource Group of the NIF Council, 4 November 1999.

Energy & Technology Review. "The National Ignition Facility: An Overview." December 1994, 1–6.

Erlandson, A. C., R. London, K. Manes, et al. *Accelerated Thermal Recovery for Flashlamp-Pumped Solid-State Laser Amplifiers*. UCRL-ID-135668. Livermore, CA: Lawrence Livermore National Laboratory, 1999.

Farber, W. R., J. C. Fraser, D. J. Grove, et al. *A Conceptual Design of the Model C Stellarator*. NYO-7309, AEC Research and Development Report. Princeton, NJ: Project Matterhorn at Princeton University, 1956.

Farnsworth, P. T. Electric Discharge Device for Producing Interactions Between Nuclei. U.S. Patent 3258402, filed 11 January 1962 and issued 28 June 1966.

——. Method and Apparatus for Producing Nuclear-Fusion Reactions. U.S. Patent 3386883, filed 13 May 1966 and issued 4 June 1968.

Feder, Toni. "DOE Warms to Cold Fusion." *Physics Today*, April 2004, 27.

Ferrell, Robert H., ed. *The Eisenhower Diaries*. New York: W. W. Norton, 1981.

Finney, John W. "A.E.C. Spurs Work on H-Bomb Power." *New York Times*, 11 February 1958.

———. "Atomic Freedom Hailed at Geneva." *New York Times*, 3 September 1958. 9.

———. "Gains in Harnessing Power of H-Bomb Reported Jointly by U.S. and Britain." *New York Times*, 25 January 1958.

———. "Navy Breakthrough in Control of Hydrogen Bomb Power Seen." *New York Times*, 18 August 1959.

———. "U.S. Experts Hint H-Bomb Control. *New York Times*, 5 September 1958.

———. "U.S. Will Show H-Power Control Devices." *New York Times*, 17 August 1958.

The Fire Place. 12 July 2007. http://fire.pppl.gov/ (accessed 20 August 2007).

Fitzpatrick, Tim. "Japan Yanks Funding for Cold-Fusion Effort." *Salt Lake Tribune*, 27 August 1997.

Fleischmann, Martin, and Stanley Pons. "Electrochemically Induced Nuclear Fusion of Deuterium." *Journal of Electroanalytical Chemistry* 261 (1989): 301–8.

———. "Some Comments on the Paper Analysis of Experiments on Calorimetry of $LiOD/D_2O$ Electrochemical Cells, R. H. Wilson et al., *J. Electroanal. Chem.* 332 (1992) 1." *Journal of Electroanalytical Chemistry* 332 (1992): 33–53.

Fleischmann, Martin, Stanley Pons, Marvin W. Anderson, Lian Jun Li, and Marvin Hawkins. "Calorimetry of the Palladium-Deuterium–Heavy Water System." *Journal of Electroanalytical Chemistry* 287 (1990): 293–348.

Fleischmann, Martin, Stanley Pons, Marvin Hawkins, and R. J. Hoffman. "Measurement of γ-rays from Cold Fusion." *Nature* 339 (29 June 1989): 667.

Fleischmann, Martin, Stanley Pons, and others. Press conference of 23 March 1989. Audio recording. Available at http://newenergytimes.com/Audio/1989UtahPC1.mp3, http://newenergytimes.com/Audio/1989UtahPC2.mp3, and http://newenergytimes.com/Audio/1989UtahPC3.mp3.

Foot, Rosemary. "Nuclear Coercion and the End of the Korean Conflict." *International Security* 13 (Winter 1998/99): 92–112.

Footlick, Jerrold K. *Truth and Consequences: How Colleges and Universities Meet Public Crises*. Phoenix: The American Council on Education and Oryx Press, 1997.

Forslund, David W., and Philip D. Goldstone. "Photon Impact: High-Energy Plasma Physics with CO_2 Lasers." *Los Alamos Science*, Spring–Summer 1985, 2–27.

Forty, C. B. A., and N. P. Taylor. "Low Activation Material Candidates for Fusion Power Plants." Paper presented at the Euromat96 Conference, Bournemouth, UK, 21–23 October 1996.

Fowler, T. Kenneth. *The Fusion Quest.* Baltimore: The Johns Hopkins University Press, 1997.

Fox, Hal. "Cold Fusion Isn't as Cold as You Think." *Deseret (Utah) Morning News,* 4 September 1997.

Frankel, Max. "Soviet Scientist Sees Candor Curb." *New York Times,* 1 March 1958.

Friedman, Edward. "Nuclear Blackmail and the End of the Korean War." *Modern China,* January 1975, 75–91.

Friendly, Alfred. "New A-Bomb Has 6 Times Power of 1st." *Washington Post,* 18 November 1949.

Fusion Power Associates. "U.S. Fails to Sign ITER Extension Agreement; JET Demonstrates Fusion Remote Maintenance." *Executive Newsletter,* August 1998.

———. "US Fusion Budgets for OFES and ICF." Available at http://aries.ucsd .edu/FPA/OFESbudget.shtml (accessed 26 July 2007).

Gallagher, Paul. "'Hit Men' vs. LaRouche's Fusion Energy Foundation." *Executive Intelligence Review,* December 3, 2004. Available at http://www.larouchepub .com/other/2004/3147_hit_men_vs_fef.html (accessed 12 July 2007).

Gamow, G., and E. Teller. "The Rate of Selective Thermonuclear Reactions." *Physical Review* 53 (April 1938): 608–9.

Garwin, Richard L. "Consensus on Cold Fusion Still Elusive." *Nature* 338 (20 April 1989): 616–17.

Gehmlich, Dietrich K. *A History of the College of Engineering, University of Utah.* 2003. http://www.coe.utah.edu/about/history.pdf (accessed 24 July 2007).

General Atomics. "The Promise of Fusion Energy." PowerPoint presentation, ca. 2004. http://fusioned.gat.com/images/pdf/promise_of_fusion.pdf (accessed 26 July 2007).

George, K. A. "Neutron Production and Temperature in ZETA." *Nature* 182 (13 September 1958): 745–46.

Gibson, A. "Possibility of Ion Runaway in ZETA." *Nature* 183 (10 January 1959): 101–2.

Gladwell, Malcolm. "The Televisionary." *New Yorker,* 27 May 2002, 112–16.

Glanz, James. "Behind the Official Optimism, Flawed Projections." *Science* 274 (6 December 1996): 1601.

———. "A Harsh Light Falls on NIF." *Science* 277 (18 July 1997): 304–7.

———. "Requiem for a Heavyweight at Meeting on Fusion Reactors." *Science* 280 (8 May 1998): 818–19.

———. "Turbulence Might Sink Titanic Reactor." *Science* 274 (6 December 1996): 1600.

Glanz, James, and Andrew Lawler. "Planning a Future Without ITER." *Science* 279 (2 January 1998): 20–21.

Goncharov, G. A. "American and Soviet H-Bomb Development Programmes: Historical Background." *Physics-Uspekhi* 39 (1996): 1033–44.

Goodchild, Peter. *Edward Teller: The Real Dr. Strangelove*. Cambridge, MA: Harvard University Press, 2004.

Gorman, J. G., I. G. Brown, G. Lisitano, and J. Orens. "New Model for Plasma Confinement Times at Stellarators." *Physical Review Letters* 22 (6 January 1969): 16–20.

Groves, Leslie R. *Now It Can Be Told*. New York: Da Capo, 1983.

Hagerman, D. C., and J. W. Mather. "Neutron Production in a High-Power Pinch Apparatus." *Nature* 181 (25 January 1958): 226–28.

Harte, J. A., W. E. Alley, D. S. Bailey, J. L. Eddlemay, and G. B. Zimmerman. "LAS-NEX—A 2-D Physics Code for Modeling ICF." *ICF Quarterly Report*. UCRL-LR-105821-96-4 (1996): 150–64.

Hattiangandi, J. N. "Alternatives and Incommensurables: The Case of Darwin and Kelvin." *Philosophy of Science* 38 (December 1971): 502–7.

Heller, Arnie. "Collaboration Ignites Laser Advances." *Science & Technology Review*, June 1999, 19–21.

———. "On Target: Designing for Ignition." *Science & Technology Review*, July 1999, 4–11.

———. "Orchestrating the World's Most Powerful Laser." *Science & Technology Review*, July–August 2005, 4–12.

Herken, Gregg. *Brotherhood of the Bomb*. New York: Henry Holt, 2002.

Herman, Robin. *Fusion: The Search for Endless Energy*. Cambridge: Cambridge University Press, 1990.

Herrmann, Mark C. "ICF Basics, NIF and IFE." PowerPoint presentation, UCRL-PRES-207384, ca. 2005. http://fire.pppl.gov/icf_hermmann_2005.pdf (accessed 26 July 2007).

Hershey, Robert D., Jr. "Nuclear Fusion Growth Aim of New Law." *New York Times*, 9 October 1980.

Hewlett, Richard G., and Jack M. Holl. *Atoms for Peace and War*. Berkeley: University of California Press, 1989.

Hillaby, John. "British Changing Fusion Machines." *New York Times*, 23 July 1959.

———. "British Chide U.S. on Nuclear Pace." *New York Times*, 17 January 1958.

———. "H-Power System to Take 20 Years." *New York Times*, 25 January 1958.

Hijiya, James A. "The *Gita* of J. Robert Oppenheimer." *Proceedings of the American Philosophical Society* 144 (June 2000): 123–67.

Hoffman, Ian. "Edward Teller Legacy Goes beyond H-Bomb." *Oakland Tribune*, 15 September 2003.

Hoffmann, M. David. *Readings for the Atomic Age*. New York: Globe, 1950.

Holloway, Rachel L. *In the Matter of J. Robert Oppenheimer: Politics, Rhetoric, and Self-Defense*. Westport, CT: Praeger, 1993.

Holtkamp, Norbert. "Status of the ITER Project." PowerPoint presentation, October 2006. http://fire.pppl.gov/aps06_holtkamp.ppt.

Honsaker, J., H. Karr, J. Osher, J. A. Phillips, and J. L. Tuck. "Neutrons From a Stabilized Toroidal Pinch." *Nature* 181 (25 January 1958): 231–33.

Hooper, E. B. *ICC 2002 Innovative Confinement Concepts*. UCRL-ID-147158-ABSTS. Livermore, CA: Lawrence Livermore National Laboratory, 2002.

Horgan, John. "Infighting Among Rival Theorists Imperils 'Hot' Fusion Lab Plan." *Scientist* 3, no. 13 (1989): 1.

Huizenga, John R. *Cold Fusion: The Scientific Fiasco of the Century*. Rochester: University of Rochester Press, 1992.

Hunt, J. T., K. R. Manes, J. R. Murray, et al. *Laser Design Basis for the National Ignition Facility*. UCRL-JC-117399. Livermore, CA: Lawrence Livermore National Laboratory, 1994.

"In the Matter of: Biophysics Conference." Transcript of "Biophysics Conference" held Tuesday, January 18, 1995 [*sic*: 1955] in Washington, DC. Available in the George Washington University National Security Archive Web site, http://www.gwu.edu/~nsarchiv/radiation/dir/mstreet/commeet/meet15/brief15/tab_d/br15d2a.txt (accessed 25 July 2007).

"In the Matter of Arbitration Between Patent Office Professional Association and U.S. Department of Commerce, Patent and Trademark Office." FMCS Case No. 00-01666; Robert T. Moore, Arbitrator. 30 July 2005. Available at http://users.erols.com/iri/ValonePatentOfficeDecision.htm (accessed 25 July 2007).

Jacquinot, J., and the JET team. "Deuterium-Tritium Operation in Magnetic Confinement Experiments: Results and Underlying Physics." *Plasma Physics and Controlled Fusion* 41 (1999): A13–A46.

Jahn, Gunnar. "Presentation Speech." Presentation of Linus Pauling, 10 December 1963. From Nobel Lectures, Peace, 1951–1970. Frederick W. Haberman, ed. Amsterdam: Elsevier, 1922. http://nobelprize.org/nobel_prizes/peace/laureates/1962/press.html.

James, William. *The Will to Believe*. New York: Longmans, Green, 1915.

Japan Atomic Energy Agency. "WWW Chart of the Nuclides 2004." 14 January 2005. http://wwwndc.jaea.go.jp/CN04/index.html (accessed 27 July 2007).

Japan Atomic Energy Research Institute. *Persistent Quest—Research Activities, 1997.* http://jolisfukyu.tokai-sc.jaea.go.jp/fukyu/tayu/ACT97E/frameset.htm (accessed 26 July 2007).

Jarvik, Elaine. "Does Fusion Scientist 'Hold the Secret?'" *Deseret (Utah) Morning News,* 24 March 2006. http://deseretnews.com/dn/view/0,1249,635194149,00 .html (accessed 24 July 2007).

JASON. *Inertial Confinement Fusion (ICF) Review.* JSR-96-300. McLean, VA: MITRE, 1996.

——. *NIF Ignition.* JSR-05-340. McLean, VA: MITRE, 2005.

——. *Science Based Stockpile Stewardship.* JSR-94-345. McLean, VA: MITRE, 1994.

Jimenez, Roman. "Argentine Tests of Atomic Power Said Completed." Associated Press, in the *Reno Evening Gazette,* 30 October 1951.

Johnston, Richard J. H. "No Danger Seen in Nuclear Tests." *New York Times,* 20 January 1956.

Jones, Richard. "Appropriators Take Issue with Administration's Nuclear Weapons Initiatives." *FYI: The AIP Bulletin of Science Policy News.* 23 May 2005.

——. "Orbach Speaks to Fusion Advisory Committee." *FYI: The AIP Bulletin of Science Policy News.* 17 March 2003.

Jones, Steven E. "Behold My Hands: Evidence for Christ's Visit in Ancient America." Early 2006. http://www.physics.byu.edu/faculty/jones/rel491/ handstext+and+figures.htm; currently defunct.

——. "Muon-Catalyzed Fusion Revisited." *Nature* 321 (8 May 1986): 127–33.

——. "Why Indeed Did the WTC Buildings Completely Collapse?" *Journal of 9-11 Studies* 3 (September 2006) http://www.journalof911studies.com/ volume/200609/Why_Indeed_Did_the_WTC_Buildings_Completely_ Collapse_Jones_Thermite_World_Trade_Center.pdf (accessed 26 July 2007).

Jones, Steven E., A. N. Anderson, A. J. Caffrey, et al. "Observation of Unexpected Density Effects in Muon-Catalyzed *d-t* Fusion." *Physical Review Letters* 56 (10 February 1986): 588–91.

Jones, Steven E., and J. Ellsworth. "Geo-fusion and Cold Nucleosynthesis." 2003. http://www.lenr-canr.org/acrobat/JonesSEgeofusiona.pdf (accessed 26 July 2007).

Jones, Steven E., E. P. Palmer, J. B. Czirr, et al. "Observation of Cold Nuclear Fusion in Condensed Matter." *Nature* 338 (27 April 1989): 737–40.

Kaempffert, Waldemar. "Argentina Lacks Development Resources Even If, in Theory, Its Atomic Tests Are Possible." *New York Times,* 1 April 1951.

Kaplan, Morris. "Fusion Reactor Gains in Britain." *New York Times,* 18 October 1957.

Kenworthy, E. W. "U.S. Thinks Soviet Will Pledge Halt In Nuclear Tests." *New York Times*, 29 March 1958.

——. "U.S. Warns Free Nations Not to Be Misled by Soviet." *New York Times*, 1 April 1958.

Konopinski, E. J., C. Marvin, and E. Teller. *Ignition of the Atmosphere with Nuclear Bombs*. LA-602. Los Alamos, NM: Los Alamos Laboratory, 1946.

Konrad, C. H. *Use of Z-Pinch Sources for High-Pressure Shock Wave Studies*. SAND98-0047. Albuquerque, NM: Sandia National Laboratories, 1998.

Krebs, Martha. Letter to ITER Council Members and IAEA Director-General. 7 October 1998.

Krivit, Steven B. "The Five Press Conferences of Cold Fusion." *NewEnergyTimes.com*, ca. 2006. http://www.newenergytimes.com/PR/TheFivePressConferences OfColdFusion.htm (accessed 26 July 2007).

Kulp, J. Laurence, and Arthur R. Schulert. "Strontium-90 in Man V." *Science* 136 (18 May 1962): 619–32.

Langmuir, Irving. "Oscillations in Ionized Gases." *Proceedings of the National Academy of Sciences* 14 (1928): 627–37.

Latham, R. "Thermonuclear Research in Great Britain." *Nature* 184 (3 October 1959): 1015–18.

Laurence, William. "Efforts to Tame Energy of H-Bomb Launch a New Field: 'Magnetohydrodynamics.'" *New York Times*, 23 November 1958.

——. "New A.E.C. Project Brings Nearer the Day of Useful Thermonuclear Power." *New York Times*, 7 April 1957.

——. "Taming of H-Bomb as Fuel Forecast Within 20 Years." *New York Times*, 9 August 1955.

——. "Nuclear Fusion Advanced by U.S., Britain Toward Creation of Unlimited Power." *New York Times*, 26 January 1958.

Leath, Audrey T. "Boehlert Demands a Plan for ITER Financing." *FYI: The AIP Bulletin of Science Policy News*. 21 November 2005.

Leviero, Anthony. "Scientists Term Radiation a Peril to Future of Man." *New York Times*, 13 June 1956.

——. "Truman Gives Aim." *New York Times*, 1 December 1950.

Lilienthal, David E. "The Case for Candor on National Security." *New York Times*, 4 October 1959.

——. *Change, Hope, and the Bomb*. Princeton: Princeton University Press, 1963.

——. *The Journals of David E. Lilienthal*. Vol. 2, *The Atomic Energy Years, 1945–1950*. New York: Harper and Row, 1964.

——. *The Journals of David E. Lilienthal*. Vol. 3, *The Venturesome Years, 1950–1955*. New York: Harper and Row, 1966.

Lindl, John. "Development of the Indirect-Drive Approach to Inertial Confinement Fusion and the Target Physics Basis for Ignition and Gain." *Physics of Plasmas* 2 (November 1995): 3933–4024.

———. *The Edward Teller Medal Lecture: The Evolution Toward Indirect Drive and Two Decades of Progress Toward ICF Ignition and Burn.* UCRL-JC-115197. Livermore, CA: Lawrence Livermore National Laboratory, 1993.

Lindley, David. "More Than Scepticism." *Nature* 339 (4 May 1989): 4.

Llewellyn-Smith, Chris, and David Ward. "Fusion Power." *European Review* 13, no. 3 (2005): 337–59.

Los Alamos Science. "The Oppenheimer Years." Winter–Spring 1983, 6–25.

Los Alamos Scientific Laboratory. "The New World." In *Los Alamos: Beginning of an Era, 1943–1945.* Los Alamos, NM: Public Relations Office, Los Alamos Scientific Laboratory, ca. 1967–71. Available at http://www.fas.org/sgp/othergov/doe/lanl/00285810.pdf.

Love, Kennett. "Britain Confirms Major Atom Gain." *New York Times*, 27 November 1957.

———. "Britain Indicates Reactor Advance." *New York Times*, 7 May 1958.

———. "British Modifying Fusion Apparatus." *New York Times*, 29 January 1958.

———. "British-U.S. Data on Hydrogen Due." *New York Times*, 13 January 1958.

———. "Briton 90% Sure Fusion Occurred." *New York Times*, 25 January 1958.

———. "Butler Affirms Atom Fusion Lead." *New York Times*, 31 January 1958.

———. "H-Bomb Untamed, Britain Admits." *New York Times*, 17 May 1958.

Macrae, Norman. *John von Neumann.* New York: Pantheon, 1992.

Maddox, John. "What to Say about Cold Fusion." *Nature* 338 (27 April 1989): 701.

Magnetic Fusion Energy Engineering Act of 1980. Public Law 96-386 (7 October 1980).

Malakoff, David. "DOE Slams Livermore for Hiding NIF Problems." *Science* 285 (10 September 1999): 1647.

Mallove, Eugene. "MIT and Cold Fusion: A Special Report." *Infinite Energy*, 1999, issue 24, 1–57.

Malik, John. *The Yields of the Hiroshima and Nagasaki Nuclear Explosions.* LA-8819. Los Alamos, NM: Los Alamos National Laboratory, 1985.

Mariscotti, Mario. *The Secret of Huemul Island.* Trans. unknown. (Unpublished draft translation of Mario Mariscotti, *Secreto Atomico de Huemul*, Buenos Aires: Sudamericana-Planeta, 1985) 1991.

Mark, Hans, and Lowell Wood. *Energy in Physics, War and Peace.* Dordrecht: Kluwer, 1988.

Mark, J. Carson. *A Short Account of Los Alamos Theoretical Work on Thermonuclear*

Weapons, 1946–1950. LA-5647-MS. Los Alamos, NM: Los Alamos Scientific Laboratory, 1974.

Mark, [J.] Carson, Raymond E. Hunter, and Jacob J. Wechsler. "Weapon Design: We've Done a Lot but We Can't Say Much." *Los Alamos Science*, Winter–Spring 1983, 159–63.

Marx, Karl. *The 18th Brumaire of Louis Napoleon.* In *The Portable Karl Marx*, trans. Eugene Kamenka. New York: Penguin, 1983.

Maugh, Thomas H., II. "College Reviews Physicist's Tabletop Fusion Claims." *Los Angeles Times*, 9 March 2006, A9.

Mayo, Santos. "More on the Value of Ronald Richter's Work." *Physics Today*, March 2004, 14.

McElheney, Victor K. "Fission, Fusion, the Sun—Energy Choices?" *New York Times*, 22 December 1976.

McMillan, Priscilla J. *The Ruin of J. Robert Oppenheimer.* New York: Viking, 2005.

McPhee, John. *The Curve of Binding Energy.* New York: Farrar, Straus and Giroux, 1973.

Ministry of Education, Culture, Sports, Science, and Technology (MEXT) of Japan. "World-Highest Plasma Performance Achieved by JT-60—New Record of Equivalent Fusion Gain of 1.25 Obtained." Press release, 25 June 1998.

Miyahara, Akira. "A Record of the First Five Meetings of 'The Symposium on Fusion Technology Held between 1960 and 1968.'" *Fusion Engineering and Design* 73 (2005): 383–94.

Morrow, Edward A. "Argentine Congress Jails Atom Expert." *New York Times*, 18 September 1954.

——. "Peronist Supreme Court Is Dissolved by Lonardi." *New York Times*, 6 October 1955.

——. "Perón's Atom Dream Fades; Director Reported Arrested." *New York Times*, 5 December 1952.

Moses, Edward I. "The Grand Challenge of Thermonuclear Ignition." *Science & Technology Review*, July–August 2005, 3.

Moss, Norman. *Men Who Play God: The Story of the H-Bomb and How the World Came to Live with It.* New York: Harper and Row, 1968.

Mueller, Marvin M. "DOE Fusion Managers and Science: A Personal Reaction." 14 June 1997. http://www.fas.org/sgp/eprint/mueller.html (accessed 10 June 2007).

Murphy, T. J. *Possible Hohlraum Geometries for Trident-Upgrade.* LA-UR-98-1503. Los Alamos, NM: Los Alamos National Laboratory, 1998.

Naranjo, B. "Comment on 'Nuclear Emissions During Self-Nucleated Acoustic

Cavitation.'" arXiv.org, physics/0603060v2, 12 September 2006. http://arXiv.org.

——. "Comment on 'Taleyarkhan et al. Reply.'" arXiv.org, physics/0702009v1, 1 February 2007. http://arXiv.org.

Naranjo, Brian, J. K. Gimzewski, and S. Putterman. "Observation of Nuclear Fusion Driven by a Pyroelectric Crystal." *Nature* 434 (28 April 2005): 1115–17.

National Nuclear Security Administration. "Defense Programs." http://www.nnsa.doe.gov/defense.htm (accessed 1 August 2007).

National Research Council. *Burning Plasma: Bringing a Star to Earth*. Washington, DC: National Academies Press, 2004.

——. *Cooperation and Competition on the Path to Fusion Energy*. Washington, DC: National Academies Press, 1984.

——. *Plasma Science: Advancing Knowledge in the National Interest*. Washington, DC: National Academies Press, 2007.

Nature. "Cold Fusion in Print." Vol. 338 (20 April 1989): 604.

——. "Controlled Nuclear Fusion Reactions." Vol. 182 (18 October 1958): 1051–53.

——. "Controlled Thermonuclear Reactions." Vol. 181 (22 March 1958): 803–6.

——. "Fuel and Power in Britain." Vol. 183 (21 February 1959): 511.

——. "Fuel Consumption and Resources." Vol. 181 (15 March 1958): 745–46.

——. "Harnessing Nuclear Fusion." Vol. 181 (25 January 1958): 213.

——. "Hopes for Nuclear Fusion Continue to Turn Cool." Vol. 338 (27 April 1989): 691.

Navratil, Gerald. "Bold Step by the World to Fusion Energy: ITER." PowerPoint presentation, 21 March 2006. http://www.apam.columbia.edu/cr2090/ConEd_Lecture.pdf (accessed 26 July 2007).

New Energy Times. "2004 U.S. Department of Energy Cold Fusion Review Reviewer Comments." http://www.newenergytimes.com/DOE/2004-DOE-ReviewerComments.pdf (accessed 26 July 2007).

New York Times. "Argentina and the Sun," 27 March 1951.

——. "Argentine Ex-Atom Head Demands Open Hearings." 10 September 1954.

——. "British Deny U.S. Gags Atomic Gain." 13 December 1957.

——. "The Cult of LaRouche." 10 October 1979.

——. "Dutch Scientist Denies Reports." 1 June 1951.

——. "Excerpts from Strauss Statement and Texts of British and A.E.C. Observations." 25 January 1958.

——. "French Atom Expert Backs Peron Claim." 1 April 1951.

——. "Fund Cuts in Study of Fusion Deplored." 27 April 1971.

——. "Fusion Gains Cited." 7 February 1958.

———. "H-Power Causes Radiation Leaks." 30 October 1958.

———. "Keeve M. Siegel, 52, Engineer, Physicist." 15 March 1975.

———. "Lilienthal Scouts Claim." 29 March 1951.

———. "Moscow's Envoy on TV in Britain." 3 February 1958.

———. "Netherlands, Argentina Near Atom Accord; Dutch Said to Offer Technical Assistance." 8 June 1951.

———. "Perón's Fizz-Bomb." 7 December 1952.

———. "'Pinch' for H-Power." 26 January 1958.

———. "Richter Reported in Buenos Aires." 12 December 1952.

———. "Scientists Show Man-Made Sun." 20 October 1958.

———. "Sic Transit Richter." 7 December 1952.

———. "Soviet Radio Hails British on Fusion." 30 January 1958.

———. "Strauss Depicts Atomic Progress." 9 January 1958.

———. "12 Scientists Ask Bomb Tests Go On." 21 October 1956.

———. "Two Texts on the Oppenheimer Case." 14 April 1954.

———. "Visit to Argentina Is Set." 17 May 1951.

Nordyke, Milo D. *The Soviet Program for Peaceful Uses of Nuclear Explosions.* UCRL-ID-124410. Livermore, CA: Lawrence Livermore National Laboratory, 1996.

"Nuclear Energy Development as a Problem of Manpower." *Nature* 182 (9 August 1958): 343–45.

Nuckolls, John. *Achieving Competitive Excellence in Nuclear Energy: The Threat of Proliferation; The Challenge of Inertial Confinement Fusion.* UCRL-JC-117385. Livermore, CA: Lawrence Livermore National Laboratory, 1994.

Nuckolls, John, Lowell Wood, Albert Theissen, and George Zimmerman. "Laser Compression of Matter to Super-High Densities: Thermonuclear (CTR) Applications." *Nature* 239 (15 September 1972): 139–42.

Oak Ridge National Laboratory. "Preliminary Evidence Suggests Possible Nuclear Emissions during Experiments." Press release, 4 March 2002.

Ogle, William E. *An Account of the Return to Nuclear Weapons Testing After the Moratorium, 1958–1961.* NVO-291. U.S. Department of Energy, Nevada Operations Office: 1985.

———. Letter to Cmdr. Russell E. Maynard, USN, 28 March 1952. Available at http://worf.eh.doe.gov/data/ihp1d/126832.pdf.

Olsen [*sic*], Thiago, and Rose Jacobs. "There's a Nuke in My Garage." *ft.com*, 26 January 2007. http://search.ft.com/ftArticle?queryText=fusor&aje=true&id=070126006711 (accessed 26 July 2007).

Olson, Thiago. "Archived—First Plasma." Online posting, 20 June 2006. http://fusor.net/board/download_thread.php?bn=fusor_images&thread=1150855195&site=fusor (accessed 26 July 2007).

O'Neill, Dan. *The Firecracker Boys.* New York: St. Martin's Griffin, 1994.

Orbach, Raymond. "Remarks of Dr. Raymond Orbach, Under Secretary for Science, U.S. Department of Energy, On the Signing of the ITER Agreement, Paris, France, November 21, 2006." http://www.iter.org/pdfs/speech_Orbach .pdf

Ord-Hume, Arthur W. J. G. *Perpetual Motion: The History of an Obsession.* Kempton, IL: Unlimited Adventures, 1977.

Osmundsen, John A. "Nuclear Experts See New Devices." *New York Times,* 14 October 1959.

Paine, Christopher. *When Peer Review Fails: The Roots of the National Ignition Facility (NIF) Debacle.* National Resources Defense Council, June 2000. http://www .nrdc.org/nuclear/nif2/nif2inx.asp.

Park, Robert. "Bubble Fusion: A Collective Groan Can Be Heard." *What's New,* 1 March 2002.

——. "A Cold Fusion Institute Was Called for in Hearings." *What's New,* 28 April 1989.

——. "The Corpse of Cold Fusion Will Probably Continue to Twitch." *What's New,* 5 May 1989.

——. "Free Energy: State Department Opens Its Doors to New Agers." *What's New,* 5 March 1999.

——. "Future Energy: New Age Energy Conference Moves to Commerce." *What's New,* 19 March 1999.

——. *Voodoo Science: The Road from Foolishness to Fraud.* New York: Oxford University Press, 2000.

Parker, Ann. "Empowering Light." *Science & Technology Review,* September–October 2002. http://www.llnl.gov/str/September02/September50th.html.

Parliamentary Office of Science and Technology. "Nuclear Fusion." *Postnote,* January 2003, no. 192.

Parrott, Lindesay. "Nuclear Downpour Hit Ship During Test at Bikini—U.S. Inquiry Asked." *New York Times,* 17 March 1954.

Pauling, Linus. "Science and Peace." Nobel Lecture, 11 December 1963. From *Nobel Lectures, Peace, 1951–1970.* Frederick W. Haberman, ed. Amsterdam: Elsevier, 1972. http://nobelprize.org/nobel_prizes/peace/laureates/1962/ pauling-lecture.html.

Payne, Stephen, and Christopher Marshall. "Taking Lasers Beyond the National Ignition Facility." *Science & Technology Review,* September 1996, 4–11.

Pease, R. S. "Experiments on the Problem of Controlled Thermonuclear Reactions." *Nature* 196 (29 December 1962): 1247–53.

Peat, F. David. *Cold Fusion: The Making of a Scientific Controversy*. Chicago: Contemporary Books, 1989.

Peplow, Mark. "Desktop Fusion is Back on the Table." *news@nature.com*, 10 January 2006. http://www.nature.com/news/2006/060109/full/060109-5.html (accessed 1 July 2007).

Peterson, Britt. "The Future of Fusion." *seedmagazine.com*, 22 June 2006. http://www.seedmagazine.com/news/2006/06/the_future_of_fusion.php (accessed 24 July 2007).

Petrasso, R. D., X. Chen, K. W. Wenzel, R. R. Parker, C. K. Li, and C. Fiore. "Petrasso et al. Reply." *Nature* (29 June 1989): 667–69.

———. "Problems with the gamma-ray Spectrum in the Fleischmann et al. Experiments." *Nature* 339 (18 May 1989): 183–85.

Phillips, James A. "Magnetic Fusion." *Los Alamos Science*, Winter–Spring 1983, 64–67.

Pitts, Richard. "Fusion: The Way Ahead?" *PhysicsWeb*, March 2006. http://physicsweb.org/articles/world/19/3/7 (accessed 26 July 2007).

Platt, Charles. "What If Cold Fusion Is Real?" *Wired*, November 1998.

Plumb, Robert. "British Industry Group Cites Fusion Advance in $28,000 Tube." *New York Times*, 25 January 1958.

Pool, Robert. "Bulls Outpace Bears—For Now!" *Science* 244 (28 April 1989): 421.

———. "Fusion Breakthrough?" *Science* 243 (31 March 1989): 1661–62.

———. "Fusion Followup: Confusion Abounds." *Science* 244 (7 April 1989): 27–29.

———. "Fusion Theories Pro and Con." *Science* 244 (21 April 1989): 285.

———. "How Cold Fusion Happened—Twice!" *Science* 244 (28 April 1989): 420–23.

———. "Skepticism Grows Over Cold Fusion." *Science* 244 (21 April 1989): 284–85.

———. "'Utah Effect' Strikes Again?" *Science* 244 (28 April 1989): 420.

———. "Wolf: My Tritium Was an Impurity." *Science* 248 (15 June 1990): 1301.

Pool, Robert, and T. A. Heppenheimer. "Electrochemists Fail to Heat Up Cold Fusion." *Science* 244 (12 May 1989): 647.

Post, R. F. "Fusion Power." *Proceedings of the National Academy of Sciences* 68 (August 1971): 1931–37.

———. "Fusion Reactors as Future Energy Sources." *Science* 186 (1 November 1974): 397–407.

Post, Richard F. "Controlled Fusion Research—An Application of the Physics of High Temperature Plasmas." *Reviews of Modern Physics* 28, no. 3 (July 1956): 338–62.

President's Committee of Advisors on Science and Technology (PCAST). *The U.S. Program of Fusion Energy Research and Development*. July 1995.

"Project Hardy (The Return of the Native.)" Los Alamos National Laboratory Records. http://worf.eh.doe.gov/data/ihp1c/0607_a.pdf (accessed 26 July 2007).

Purdue University. "Evidence Bubbles Over to Support Tabletop Nuclear Fusion Device." Press release, 2 March 2004.

——. "Panel Determines Research Allegations Merit Additional Investigation." Press release, 10 September 2007.

——. "Purdue Findings Support Earlier Nuclear Fusion Experiments." Press release, 12 July 2005.

——. "Purdue Integrity Panel Completes Research Inquiry." Press release, 7 February 2007.

——. "Report of the Investigation Committee In the Matter of Rusi P. Taleyarkhan." Final Report of the C-22 Investigation Committee, 18 April 2008.

Putterman, Seth J. "Sonoluminescence: Sound into Light." *Scientific American,* February 1995, 32–37.

——. "Sonoluminescence: The Star in a Jar." *Physics World,* May 1998, 38–42.

Putterman, S. J., L. A. Crum, and K. Suslick. "Comments on 'Evidence for Nuclear Emissions During Acoustic Cavitation'; by R. P. Taleyarkhan et al., *Science* 295, 1868, March 8, 2002." www.arXiv.org, cond-mat/0204065v2, 2 April 2002. http://arxiv.org.

Rafelski, Johann, and Steven E. Jones. "Cold Nuclear Fusion." *Scientific American,* July 1987, 84–89.

Reams, C. Austin. *Russia's Atomic Tsar: Viktor N. Mikhailov.* LA-UR-97-234. Los Alamos, NM: Center for International Security Affairs at Los Alamos National Laboratory, 1996.

Reich, Eugenie Samuel. "Bubble Bursts for Table-Top Fusion." *news@nature.com,* 8 March 2006. http://www.nature.com/news/2006/060306/full/060306-3.html (accessed 1 July 2007).

——. "Bubble-Fusion Group Suffer Setback." *news@nature.com,* 10 May 2006. http://www.nature.com/news/2006/060508/full/060508-8.html (accessed 1 July 2007).

——. "Bubble Fusion: Silencing the Hype." *news@nature.com,* 8 March 2006. http://www.nature.com/news/2006/060306/full/060306-1.html (accessed 1 July 2007).

——. "Concerns Grow over Secrecy of Bubble-Fusion Inquiry." *Nature* 442 (20 July 2006): 230–31.

——. "Congress Requests Bubble-Fusion Reports." *news@nature.com,* 28 March 2007. http://www.nature.com/news/2007/070326/full/446480a.html (accessed 1 July 2007).

——. "Disputed Inquiry Clears Bubble-Fusion Engineer." *news@nature.com,* 13

February 2007. http://www.nature.com/news/2007/070212/full/445690a. html (accessed 1 July 2007).

——. "Evidence for Bubble Fusion Called into Question." *Nature* 440 (9 March 2006): 132.

——. "Is Bubble Fusion Simply Hot Air?" *news@nature.com,* 8 March 2006. http:// www.nature.com/news/2006/060306/full/060306-2.html (accessed 1 July 2007).

——. "Purdue Attacked over Fusion Inquiry." *Nature* 444 (7 December 2006): 664–65.

——. "Purdue Dogged by Misconduct Claims." *news@nature.com,* 15 May 2007. http://www.nature.com/news/2007/070514/full/447238a.html (accessed 1 July 2007).

——. "A Sound Investment?" *news@nature.com,* 8 March 2006. http://www .nature.com/news/2006/060306/full/060306-4.html (accessed 1 July 2007).

Reinhold, Robert. "More Must Share the Research Pie." *New York Times,* 30 December 1979.

Reines, F., and B. R. Suydam. *Preliminary Survey of Physical Effects Produced by a Super Bomb.* LAMS-993. Los Alamos, NM: Los Alamos Scientific Laboratory, 1949.

Reines, F., and T. J. White. *Yield of the Hiroshima Bomb.* LA-1398. Los Alamos, NM: Los Alamos Scientific Laboratory, 1952.

Reston, James. "Confusion on Atom Tests." *New York Times,* 28 March 1958.

——. "Dr. Oppenheimer Suspended By A.E.C. in Security Review; Scientist Defends Record." *New York Times,* 13 April 1954.

——. "The H-Bomb Decision." *New York Times,* 8 April 1954.

——. "The 'Low-Down' on that 'High-Up' Official." *New York Times,* 18 April 1954.

——. "Science, Education, and the Arts of Peace." *New York Times,* 25 April 1954.

Reuters. "Fusion Device Built." *New York Times,* 25 January 1958.

——. "New Fusion Tests Told." *New York Times,* 15 June 1958.

——. "Perón to Run Atom Unit." *New York Times,* 23 May 1951.

——. "Richter Ouster Reported." *New York Times,* 18 December 1952.

Rhodes, Richard. *Dark Sun: The Making of the Hydrogen Bomb.* New York: Touchstone, 1996.

——. *The Making of the Atomic Bomb.* New York: Simon and Schuster, 1986.

Richter, Ronald. "Argentina Has No Atom Bomb." *United Nations World,* July 1951, i–iii.

Roederer, Juan G. "Roederer Replies:" *Physics Today*, August 2003, 12.

Rose, B., A. E. Taylor, and E. Wood. "Measurement of the Neutron Spectrum from ZETA." *Nature* 181 (14 June 1958): 1630–32.

Rosenthal, A. M. "U.N. Circles Wary on Atom Bomb Use." *New York Times*, 1 December 1950.

Ryder, Todd H. "Fundamental Limitations on Plasma Fusion Systems Not in Thermodynamic Equilibrium." *Physics of Plasmas* 4 (April 1997): 1039–46.

Safe Energy Communication Council. "Congress Phases Out International Fusion Project Funding." Press release, 7 October 1998.

Sakharov, Andrei. *Memoirs.* Trans. Richard Lourie. New York: Alfred A. Knopf, 1990.

Samm, Ulrich. "Controlled Thermonuclear Fusion Enters with ITER into a New Era." Ca. 2003. http://www.jet.efda.org/documents/articles/samm.pdf (accessed 26 July 2007).

Sanders, Ralph. "Defense of Project Plowshare." *Technology and Culture* 4 (Spring 1963): 252–55.

Sandia National Laboratories. "Project Plowshare." Poster. http://www.sandia.gov/recordsmgmt/exhibits/PlowshareProgram.pdf (accessed 25 July 2007).

———. "Rapid-Fire Pulse Brings Sandia Z Method Closer to Goal of High-Yield Fusion Reactor." Press release, 24 April 2007.

Sauthoff, Ned. "US ITER Project." PowerPoint presentation, 31 October 2006. http://fire.pppl.gov/aps06_iter_sauthoff.pdf (accessed 26 July 2007).

Schirber, Michael. "For Nuclear Fusion, Could Two Lasers Be Better Than One?" *Science* 310 (9 December 2005): 1610–11.

Schmeck, Harold M. "Nuclear Fusion Reported in Lab with Aid of Laser." *New York Times*, 14 May 1974.

Schmidt, Dana Adams. "Soviet Gives the U.S. a Lesson in Propaganda." *New York Times*, 30 March 1958.

Schwartz, Harry. "Toward Controlled Fusion Power." *New York Times,* 26 October 1969.

"Science Museum Exhibit on Controlled Nuclear Fusion." *Nature* 182 (20 December 1958): 1710.

Seaborg, Glenn, with Benjamin S. Loeb. *Stemming the Tide: Arms Control in the Johnson Years.* Lexington, MA: Lexington, 1987.

Seife, Charles. "At the Going Down of the Nuclear Sun." *Economist,* 16 September 1995, 93–96.

———. "'Bubble Fusion' Paper Generates Tempest in a Beaker." *Science* (8 March 2002): 1808–9.

——. "CERN's Gamble Shows Perils, Rewards of Playing the Odds." *Science* 286 (29 September 2000): 2260–62.

——. "Chemistry Casts Doubt on Bubble Reactions." *ScienceNOW*, 24 July 2002. http://sciencenow.sciencemag.org/cgi/content/full/2002/724/2 (accessed 30 June 2007).

——. "Fusion Catches a Cold." *New Scientist*, 17 October 1998, 4.

——. "Honey, I Shrank the Reactor." *New Scientist*, 3 April 1999, 66.

——. "Stunning Gun." *New Scientist*, 12 June 1999, 10.

——. "Tabletop Accelerator Breaks 'Cold Fusion' Jinx But Won't Yield Energy, Physicists Say." *Science* 308 (29 April 2005): 613.

——. "Will NIF Live Up to Its Name?" *Science* 289 (18 August 2000): 1128.

Serber, Robert. *The Los Alamos Primer*. Berkeley: University of California Press, 1992.

Shabad, Theodore. "Soviet Reports a Major Step Toward a Fusion Plant." *New York Times*, 20 October 1979.

Sheffield, John. *Report on Fusion Energy Workshop, August 13–14, 1998, Washington, DC*. Knoxville: The Joint Institute for Energy and Environment, 1998.

Shepley, James R., and Clay Blair, Jr. *The Hydrogen Bomb: The Men, the Menace, the Mechanism*. New York: David McKay, 1954.

Shipman, Thomas L. (Los Alamos Scientific Lab.) "Following information is boiled down version. . . ." Telegram to John C. Bucher, Director, Division of Biology and Medicine (U.S.A.E.C.), 9 March 1954. Available at http://worf.eh.doe.gov/data/ihp1d/400079e.pdf.

Siegel, Lee. "America Condemns Inventive Science, Fathers of Cold Fusion Tell ABC." *Salt Lake Tribune*, 1 June 1994.

Smith, R. Jeffrey. "Lab Officials Squabble Over X-ray Laser." *Science* 230 (22 November 1985): 923.

Smyth, Henry de Wolf. *Atomic Energy for Military Purposes*. 1 July 1945. Available at http://www.atomicarchive.com/Docs/SmythReport/index.shtml.

Spielman, R. B., G. T. Baldwin, and G. Cooper. *D-D Fusion Experiments Using Fast Z Pinches*. SAND98-0705. Albuquerque, NM: Sandia National Laboratories, 1998.

Spitzer, Lyman. "Co-operative Phenomena in Hot Plasmas." *Nature* 181 (25 January 1958): 221–22.

Spitzer, Lyman, Jr., D. J. Grove, W. E. Johnson, L. Tonks, and W. F. Westendorp. *Problems of the Stellarator as a Useful Power Source*. NYO-6047, Princeton, NJ: Project Matterhorn at Princeton University, 1954.

Stopinski, Orin W. *Fallout Prediction as of 1957*. LA-9993-MS. Los Alamos, NM: Los Alamos Scientific Laboratory, ca. 1957.

Storm, E., S. H. Batha, T. P. Bernat, et al. *The LLNL ICF Program: Progress Toward Ignition and Gain in the Laboratory*. UCRL-JC-106080. Livermore, CA: Lawrence Livermore National Laboratory, 1990.

Sullivan, J. A., and D. B. Harris. *Development of KrF Lasers for Inertial Confinement Fusion*. LA-UR-91-7. Los Alamos, NM: Los Alamos National Laboratory, 1991.

Sullivan, Walter. "Delay in Fusion Tests Predicted." *New York Times*, 26 February 1977.

———. "First Credit Given to Small Company for Gain in Laser Fusion Power." *New York Times*, 27 April 1975.

———. "Fusion: The Answer to Fission?" *New York Times*, 15 May 1979.

———. "Gain Is Foreseen in Controlling H-Bomb Energy in 3 to 5 Years." *New York Times*, 4 February 1968.

———. "Limitless Atomic Power Sought in U. of Rochester Laser Project." *New York Times*, 5 March 1976.

———. "Nations Plan an Experimental Fusion Power Plant." *New York Times*, 21 January 1979.

———. "New Devices Mimic Solar Fusion Action." *New York Times*, 11 December 1977.

———. "New Fusion Device Hailed for Its Energy Potential." *New York Times*, 1 June 1980.

———. "New Fusion Effort Reported by Soviet." *New York Times*, 23 July 1977.

———. "New Fusion Reactor Likely to Break Even in Fuel Use." *New York Times*, 22 March 1976.

———. "Reports Assay Failures and Hopes for Fusion Power." *New York Times*, 13 January 1980.

———. "Scientists Achieve Fusion Reaction by Firing an Electron Beam at Fuel." *New York Times*, 11 June 1977.

———. "Two Laser Projects Take Scientists Toward Harnessing Stars' Energy." *New York Times*, 29 April 1978.

"Summaries of Addresses of Presidents of Sections." *Nature* 182 (supplement, 30 August 1958): 579.

Taleyarkhan, R. P., R. C. Block, R. T. Lahey, R. I. Nigmatulin, and Y. Xu. "Taleyarkhan et al. Reply:" *Physical Review Letters* 97 (6 October 2006): 149404.

Taleyarkhan, R. P., C. D. West, R. T. Lahey, R. I. Nigmatulin, R. C. Block and Y. Xu. "Nuclear Emissions During Self-Nucleated Acoustic Cavitation." *Physical Review Letters* 96 (27 January 2006): 034301-1–034301-4.

Taubes, Gary. *Bad Science: The Short Life and Weird Times of Cold Fusion*. New York: Random House, 1993.

———. "Cold Fusion Conundrum at Texas A&M." *Science* 248 (15 June 1990): 1299–1304.

Taylor, N. P., C. B. A. Forty, D. A. Petti, and K. A. McCarthy. "The Impact of Materials Selection on Long-Term Activation in Fusion Power Plants." Paper presented at The Ninth International Conference on Fusion Reactor Materials (ICFRM-9), Colorado Springs, 10–15 October 1999.

Taylor, Theodore B. "Nuclear Weapons Responsibility." Presentation at Mickleton (NJ) Monthly Meeting, Religious Society of Friends (Quakers), 20 April 1998. Transcript available at http://www.sondra.net/concerns/ttspeech.htm (accessed 25 July 2007).

Teller, Edward. *Energy from Heaven and Earth*. San Francisco: W. H. Freeman, 1979.

———. "Seven Hours of Reminiscences." *Los Alamos Science*, Winter–Spring 1983, 190–95.

Teller, Edward, and Allen Brown. *The Legacy of Hiroshima*. Garden City, NY: Doubleday, 1962.

Teller, Edward, Wilson K. Talley, and Gary H. Higgins. *The Constructive Uses of Nuclear Explosives*. New York: McGraw-Hill, 1968.

Teller, Edward, with Judith Shoolery. *Memoirs: A Twentieth-Century Journey in Science and Politics*. New York: Basic Books, 2002.

Thirring, Hans. "Is Perón's A-Bomb a Swindle?" *United Nations World*, May 1951, 1–2.

Thomson, William. "On the Age of the Sun's Heat." In *Popular Lectures and Addresses*. Vol. 1, *Constitution of Matter*. 2d ed. London and New York: Macmillan, 1891. Available at http://books.google.com/books?id=JcMKAAAAIAAJ&pg=PA356&lpg=PA356.

———. "On the Origin of Life." In *Report of the Forty-first Meeting of the British Association for the Advancement of Science; held at Edinburgh in August 1871*. London: John Murray, 1872. Available at http://books.google.com/books?id=5PI4AAAAMAAJ.

———. "On the Secular Cooling of the Earth." In *Mathematical and Physical Papers*. Vol. 3, *Elasticity, Heat, Electro-Magnetism*. London: C. J. Clay and Sons, 1890. Available at http://books.google.com/books?id=xrEEAAAAMAAJ&pg=PA295.

Thonemann, P. C., E. P. Butt, R. Carruthers, et al. "Production of High Temperatures and Nuclear Reactions in a Gas Discharge." *Nature* 181 (25 January 1958): 217–20.

Time. "The Atomic Future." 22 August 1959.

———. "A Chronology of Nuclear Confusion." 8 May 1989.

———. "Controlled Fusion." 25 July 1955.

——. "The Decision L Is Yes." 6 February 1950.

——. "Dictatorship & Corruption." 30 April 1956.

——. "Double Check." 04 June 1951.

——. "A Doughnut for Power." 17 February 1975.

——. "Energy of the Pampas." 16 April 1951.

——. "Getting Closer." 31 August 1959.

——. "Knowledge Is Power." 18 November 1957.

——. "Magnetic Bottle." 18 June 1956.

——. "Monster Conference." 15 September 1958.

——. "New Nuclear Energy?" 7 January 1957.

——. "Soviet-Controlled Fusion." 7 May 1956.

——. "Sudden Zeus." 28 March 1960.

——. "Toward H-Power." 3 February 1958.

Toth, Robert C. "Teller Opposes Test Ban Treaty." *New York Times*, 15 August 1963.

——. "Teller Shows Consistency in Opposing Test Ban." *New York Times*, 25 August 1963.

"Treaty Banning Nuclear Weapon Tests in the Atmosphere, in Outer Space and Under Water." Signed 5 August 1963. http://www.state.gov/t/ac/trt/4797 .htm.

Truman, Harold S. "President Truman's Report to the Nation on the Potsdam Conference." White House News Release, 9 August 1945. http://www.ibiblio .org/pha/policy/1945/450809a.html (accessed 26 July 2007).

——. "Statement by President Truman, September 23, 1949." 23 September 1949. Available at http://www.yale.edu/lawweb/avalon/decade/decad244.htm.

Trumbull, Robert. "Japan Achieves Nuclear Fusion." *New York Times,* 9 February 1958.

Ulam, Stanislaw M. *Adventures of a Mathematician.* New York: Charles Scribner's Sons, 1976.

United Press. "Argentina Is Claiming Short Cut to H-Bomb." *Statesville (North Carolina) Daily Record,* 29 March 1951.

——. "Argentine Hedges on Atomic Claim." *New York Times*, 29 March 1951.

——. "Britons Report Gain on Fusion Reaction." *New York Times,* 18 November 1957.

——. "Moscow Takes a Bow." *New York Times,* 25 January 1958.

——. "Perón Decorates Atomic Aid." *New York Times,* 29 March 1951.

——. "Strauss' View Derided." *New York Times,* 29 August 1956.

United Press International. "Michigan Teen Creates Nuclear Fusion." 19 November 2006.

——. "$20 Billion Voted for Nuclear Fusion." *New York Times*, 26 August 1980.

United States Atomic Energy Commission. *In the Matter of J. Robert Oppenheimer: Transcript of Hearing before Personnel Security Board and Texts of Principal Documents and Letters*. Cambridge, MA: MIT Press, 1970.

U.S. Congress, House Committee on Science and Technology, Subcommittee on Investigations and Oversight, Chairman Brad Miller. Letter to Martin C. Jischke, 21 March 2007. http://democrats.science.house.gov/Media/File/Reports/purdue_staff_report.pdf.

——, House Committee on Science and Technology, Subcommittee on Investigations and Oversight, Chairman Brad Miller. Letter to Martin C. Jischke, 9 May 2007. http://democrats.science.house.gov/Media/File/ForReleases/miller_purdue_07may09.pdf.

——, House Committee on Science and Technology, Subcommittee on Investigations and Oversight, Staff. "RE: Investigation by Purdue University of Allegations of Research Misconduct against Dr. Rusi Taleyarkhan." Memo to Rep. Brad Miller, 7 May 2007. http://democrats.science.house.gov/Media/File/AdminLetters/miller_purdue_taleyarkhan.pdf.

——, Office of Technology Assessment. *The Fusion Energy Program: The Role of TPX and Alternate Concepts*. OTA-BP-ETI-141. Washington, DC: U.S. Government Printing Office, 1995.

——, Office of Technology Assessment. *Starpower: The U.S. and the International Quest for Fusion Energy*. OTA-E-338. Washington, DC: U.S. Government Printing Office, 1987.

U.S. Department of Defense, Defense Nuclear Agency. *Castle Series, 1954*. DNA 6035F. April 1982.

——, Defense Nuclear Agency. *Operation Greenhouse, 1951*. DNA 6034F. June 1983.

——, Defense Nuclear Agency. *Operation Ivy, 1952*. DNA 6036F. December 1982.

——, Defense Nuclear Agency. *Plumbbob Series, 1957*. DNA 6005F. September 1981.

——, Defense Nuclear Agency. *Projects Gnome and Sedan—The Plowshare Program*. DNA 6029F. March 1983.

——, Defense Nuclear Agency. *Safety Experiments, November 1955–March 1958*. DNA 6030F. August 1982.

——, Defense Nuclear Agency. *Shot Apple2—A Test of the TEAPOT Series, 5 May 1955*. DNA 6012F. November 1981.

——, Defense Nuclear Agency. *Shot Bee—A Test of the TEAPOT Series, 22 March 1955*. DNA 6011F. November 1981.

——, Defense Nuclear Agency. *Shots Ess through Met and Shot Zucchini—The Final TEAPOT Tests, 23 March–15 May 1955*. DNA 6013F. November 1981.

——, Defense Nuclear Agency. *Shots Wasp through Hornet—The First Five TEAPOT Tests, 18 February–12 March 1955*. DNA 6010F. November 1981.

——, Defense Nuclear Agency. *Shots Wheeler to Morgan—The Final Eight Tests of the PLUMBBOB Series, 6 September–7 October 1957*. DNA 6007F. September 1981.

U.S. Department of Energy. *Department of Energy Assessment of the ITER Project Cost Estimate*. November 2002.

——, Energy Research Advisory Board. *Cold Fusion Research*. DOE/S-0073. November 1989.

——, Nevada Operations Office. *United States Nuclear Tests: July 1945 through September 1992*. DOE/NV-209-REV-15. December 2000.

——, Office of Declassification. *Draft Public Guidelines to Department of Energy Classification of Information*. 27 June 1994. http://www.osti.gov/opennet/forms .jsp?formurl=document/guidline/pubg.html#ZZ0 (accessed 26 July 2007).

——, Office of Energy Research. *Fusion Energy Sciences Advisory Committee: Advice and Recommendations to the U.S. Department of Energy*. DOE/ER-0690. July 1996.

——, Office of Energy Research. *Report of the FESAC/IFE Review Panel*. July 1996.

——, Office of Energy Research. *Report of FESAC Panel*. January 1998.

——, Office of Energy Research. *Strategic Plan for the Restructured Energy Sciences Program*. DOE/ER-0684. August 1996.

——, Office of Energy Research. *Fusion Power Advisory Committee (FPAC) Final Report*. DOE/S-0081. September 1990.

——, Office of Science. *Fusion Energy Sciences Advisory Committee: Review of Burning Plasma Physics*. DOE/SC-0041. September 2001.

——, Office of Science. *Fusion Energy Sciences Advisory Committee: Review of the Inertial Fusion Energy Program*. DOE/SC-0087. March 2004.

——, Office of Science. *Opportunities in the Fusion Energy Sciences Program*. June 1999.

——, Office of Science. *Report of the Review of Low Energy Nuclear Reactions*. December 2004.

——, Secretary of Energy Advisory Board. *Realizing the Promise of Fusion Energy: Final Report on the Task Force on Fusion Energy*. August 1999.

U.S. Department of State. *Foreign Relations of the United States, 1964–1968*. Vol. 11, Arms Control and Disarmament. http://www.state.gov/www/about_state/ history/vol_xi/g.html (accessed 26 July 2007).

U.S. Library of Congress, Congressional Research Service. *Congress and the Fusion Energy Sciences Program: A Historical Analysis*, by Richard E. Rowberg. January 2000.

——, Congressional Research Service. *The National Ignition Facility and Stockpile Stewardship*, by Richard E. Rowberg. March 1997.

University of California, Lawrence Livermore National Laboratory. *Laser Programs: The First 25 Years . . . 1972–1997*. URCL-TB-128043. Washington, DC: National Technical Information Service.

Urey, Harold C., F. G. Brickwedde, and G. M. Murphy. "A Hydrogen Isotope of Mass 2 and Its Concentration." *Physical Review* 40 (April 1932): 1–15.

Valiunas, Algis. "The Agony of Atomic Genius." *New Atlantis*, Fall 2006, 85–104.

Vance, Erik. "The Bursting of Bubble Fusion." *Chronicle of Higher Education*, 6 April 2007.

Vedantam, Shankar. "Fusion Experiment Sparks an Academic Brawl." *Washington Post*, 11 March 2002.

——. "'Tabletop' Fusion Report Elicits Mixed Reaction. *Washington Post*, 5 March 2002.

Vergano, Dan. "Fusion in a Beaker? Report Leaves Some Physicists Cold." *USA-Today*, 5 March 2002.

Voss, David. "A Free Energy Enthusiast Seeks Like-Minded Colleagues." *Science* 284 (21 May 1999): 1254.

Walch, Tad. "BYU Professor in Dispute over 9/11 Will Retire." *Deseret (Utah) Morning News,* 22 October 2006. http://deseretnews.com/dn/view/0,1249,650200587,00.html (accessed 26 July 2007).

Waldrop, M. Mitchell. "Cold Water from Caltech." *Science* 244 (5 May 1989): 523.

Wallheimer, Brian. "Purdue Letter That Initially Cleared Embattled Researcher Surfaces." *(Lafayette, Indiana) Journal and Courier*, 18 May 2007. http://www.jconline.com/apps/pbcs.dll/article?AID=200770518040 (accessed 1 July 2007; currently defunct).

Wall Street Journal. "Fusing the Impossible." 12 April 1989.

Warren, Virginia Lee. "Perón Announces New Way to Make Atom Yield Power." *New York Times*, 25 March 1951.

——. "Peron Is Scornful of Atomic Skeptics." *New York Times*, 26 March 1951.

Weisberg, Jacob. "Dies Ira." *New Republic*, 24 January 1994, 18–24.

Wells, Ken. "For Two Scientists, Fusion Creates Fame." *Wall Street Journal,* 27 March 1989.

Wilber, Donald N. "Overthrow of Premier Mossadeq of Iraq." Central Intelligence Agency, Clandestine Service Historical Paper No. 208, March 1954. Available at http://www.nytimes.com/library/world/mideast/iran-cia-intro.pdf.

Wilson, George. *Religio Chemici*. London: Macmillan, 1862.

Wilson R. H, J. W. Bray, P. G. Kosky, H. B. Vakil, and F. G. Will. "Analysis of Experiments on Calorimetry of LiOD-D_2O Electrochemical Cells." *Journal of Electroanalytical Chemistry* 332 (1992): 1–31.

Winterberg, Friedwardt. "Ronald Richter: Genius or Nut?" *Physics Today*, August 2003, 12.

——. "Winterberg Replies:" *Physics Today*, March 2004, 14.

Wolf, Hugh C. "The H-Bomb: Facts & Illusions." *United Nations World*, August 1951, i–iv.

Wolff, Walter. "A Typical Los Alamos National Laboratory Underground Nuclear Test." *Los Alamos Mini-Review*, September 1984. LALP-84–47 Rev.

Xu, Yiban, and Adam Butt. "Confirmatory Experiments for Nuclear Emissions During Acoustic Cavitation." *Nuclear Engineering and Design* 235 (2005): 1317–24.

Yoder, Stephen Kreider. "Japan's Scientists Join Race to Verify Fusion Experiment at Utah University." *Wall Street Journal*, 18 April 1989.

INDEX

Page numbers in *italics* refer to illustration captions.

FOR MORE FROM CHARLES SEIFE, LOOK FOR THE

Alpha and Omega
The Search for the Beginning and End of the Universe

Alpha and Omega is a dispatch from the front lines of the cosmological revolution that is being waged at observatories and laboratories around the world—in Europe, in America, and even in Antarctica—where scientists are actually peering into both the cradle of the universe and its grave. Scientists, on the trail of dark matter, dark energy, and the growing inhabitants of the particle zoo, now know how the universe will end and are on the brink of understanding its beginning. Seife makes cutting-edge science both accessible and wonderfully exciting.

ISBN 978-0-14-200446-3

Decoding the Universe
How the New Science of Information Is Explaining Everything in the Cosmos, from Our Brains to Black Holes

Previously the domain of philosophers and linguists, information theory has now moved beyond the province of code breakers to become the crucial science of our time. In *Decoding the Universe*, Charles Seife draws on his gift for making cutting-edge science accessible to explain how this new tool is deciphering everything from the purpose of our DNA to the parallel universes of our Byzantine cosmos. The result is an exhilarating adventure that deftly combines cryptology, physics, biology, and mathematics to cast light on the new understanding of the laws that govern life and the universe. *ISBN 978-0-14-303839-9*

Zero
The Biography of a Dangerous Idea

The Babylonians invented it, the Greeks banned it, the Hindus worshipped it, and the Church used it to fend off heretics. Now it threatens the foundations of modern physics. For centuries the power of zero savored of the demonic; once harnessed, it became the most important tool in mathematics. For zero, infinity's twin, is not like other numbers. It is both nothing and everything. In *Zero*, Charles Seife follows this innocent-looking number from its birth as an Eastern philosophical concept to its struggle for acceptance in Europe, its rise and transcendence in the West, and its ever-present threat to modern physics.
Winner of the PEN/Martha Albrand Award *ISBN 978-0-14-029647-1*